Industrial Strategies and Solutions for 3D Printing

Industrial Strategies and Solutions for 3D Printing

Applications and Optimization

Edited by

Hamid Reza Vanaei
Ecole Supérieure d'Ingénieurs Léonard de Vinci (ESILV)
Léonard de Vinci University
Paris, France

Sofiane Khelladi
Arts et Métiers Institute of Technology
Paris, France

Abbas Tcharkhtchi
Arts et Métiers Institute of Technology
Paris, France

Published by John Wiley & Sons, Inc., Hoboken, New Jersey.
Published simultaneously in Canada.

For general information on our other products and services or for technical support, please contact our Customer Care Department within the United States at (800) 762-2974, outside the United States at (317) 572-3993 or fax (317) 572-4002.

Wiley also publishes its books in a variety of electronic formats. Some content that appears in print may not be available in electronic formats. For more information about Wiley products, visit our web site at www.wiley.com.

Library of Congress Cataloging-in-Publication Data applied for:

Hardback ISBN: 9781394150304

Cover Design: Wiley
Cover Image: © Xuanyu Han/Getty Images; Courtesy Hamid Reza Vanaei

Set in 9.5/12.5pt STIXTwoText by Straive, Chennai, India

Contents

List of Contributors

Sidonie F. Costa
Centre for Research and Innovation in Business Sciences and Information Systems
School of Management and Technology (ESTGF)
Porto Polytechnic Institute (IPP)
Felgueiras
Portugal

José A. Covas
Institute for Polymers and Composites (IPC)
University of Minho
Guimarães
Portugal

Michael Deligant
Arts et Métiers Institute of Technology
CNAM
LIFSE
HESAM University
75013
Paris
France

Nicholas J. Dunne
School of Mechanical and Manufacturing Engineering
Dublin City University
Dublin
Ireland

and

Centre for Medical Engineering Research
School of Mechanical and Manufacturing Engineering
Dublin City University
Dublin
Ireland

and

Advanced Manufacturing Research Centre (I-Form)
School of Mechanical and Manufacturing Engineering
Dublin City University
Dublin
Ireland

and

Biodesign Europe
Dublin City University
Dublin
Ireland

and

Advanced Processing Technology Research Centre
Dublin City University
Dublin
Ireland

and

School of Pharmacy
Queen's University Belfast
Belfast
UK

and

School of Chemical Science
Dublin City University
Dublin
Ireland

and

Advanced Materials and Bioengineering Research Centre (AMBER)
Trinity College Dublin
Dublin
Ireland

and

Department of Mechanical and Manufacturing Engineering
School of Engineering
Trinity College Dublin
Dublin
Ireland

and

Trinity Centre for Biomedical Engineering
Trinity Biomedical Sciences Institute
Trinity College Dublin
Dublin
Ireland

Fernando M. Duarte
Institute for Polymers and Composites (IPC)
University of Minho
Guimarães
Portugal

Anouar El Magri
Euromed Polytechnic School
Euromed Research Center
Euromed University of Fes
Fès
Morocco

Mohammad Elahinia
Department of Mechanical Industrial and Manufacturing Engineering
University of Toledo
Toledo
OH
43606
USA

Abdulmajid Eslami
Department of Materials Engineering
Isfahan University of Technology
Isfahan
Iran

Reza Eslami Farsani
Faculty of Mechanical Engineering
K. N. Toosi University of Technology
Tehran
Iran

Shahriar Hashemipour
Department of Material Engineering
Iran University of Science and Technology
Tehran
Iran

Wuzhen Huang
School of System Design and Intelligent Manufacturing
Southern University of Science and Technology
Shenzhen
PR China

Hamid Reza Javadinejad
Department of Materials Engineering
Isfahan University of Technology
Isfahan
Iran

Namhun Kim
Department of Mechanical Engineering
Ulsan National Institute of Science and Technology
Ulsan
Republic of Korea

Sofiane Khelladi
Arts et Métiers Institute of Technology
CNAM
LIFSE
HESAM University
75013
Paris
France

Tanya J. Levingstone
School of Mechanical and Manufacturing Engineering
Dublin City University
Dublin
Ireland

and

Centre for Medical Engineering Research
School of Mechanical and Manufacturing Engineering
Dublin City University
Dublin
Ireland

and

Advanced Manufacturing Research Centre (I-Form)
School of Mechanical and Manufacturing Engineering
Dublin City University
Dublin
Ireland

and

Biodesign Europe
Dublin City University
Dublin
Ireland

and

Advanced Processing Technology
Research Centre
Dublin City University
Dublin
Ireland

Xiaofan Luo
Polymaker LLC
Houston
TX
USA

Amrid Mammeri
Arts et Métiers Institute of
Technology
CNAM
LIFSE
HESAM University
Paris
France

and

Valeo Thermal Systems
La verriere-Paris
France

Mohammad Moezzibadi
Arts et Metiers Institute of
Technology
CNAM
LIFSE
HESAM University
75013
Paris
France

Zohreh Mousavi Nejad
School of Mechanical and
Manufacturing Engineering
Dublin City University
Dublin
Ireland

and

Centre for Medical Engineering
Research
School of Mechanical and
Manufacturing Engineering
Dublin City University
Dublin
Ireland

Kaddour Raissi
Arts et Métiers Institute of
Technology
CNAM
LIFSE
HESAM University
75013
Paris
France

Kasin Ransikarbum
Department of Industrial
Engineering
Ubonratchathani University
Ubonratchathani
Thailand

and

Department of Industrial
Engineering
Kasetsart University
Bangkok
Thailand

Mohammadali Rastak
Department of Mechanical
Industrial and Aerospace
Engineering
Concordia University
Montreal
Quebec
Canada

Abbas Tcharkhtchi
Arts et Métiers Institute of
Technology
CNRS
CNAM
PIMM
HESAM University
75013
Paris
France

Hamid Reza Vanaei
ESILV
Léonard de Vinci Pôle Universitaire
92916
Paris La Défense
France

and

Léonard de Vinci Pôle Universitaire
Research Center
92916
Paris La Défense
France

and

Arts et Métiers Institute of
Technology
CNAM
LIFSE
HESAM University
75013
Paris
France

Saeedeh Vanaei
Department of Mechanical
Industrial and Manufacturing
Engineering
University of Toledo
Toledo
OH
43606
USA

Shohreh Vanaei
Department of Bioengineering
Northeastern University
Boston
MA
USA

Sébastien Vaudreuil
Euromed Polytechnic School
Euromed Research Center
Euromed University of Fes
Fès
Morocco

Yi Xiong
School of System Design and
Intelligent Manufacturing
Southern University of Science and
Technology
Shenzhen
PR China

Mengxue Yan
School of Mechanical Engineering
Changshu Institute of Technology
Changshu
Suzhou
Jiangsu
China

Nader Zirak
Arts et Métiers Institute of Technology
CNRS
CNAM
PIMM
HESAM University
75013
Paris
France

Preface

In the ever-evolving landscape of manufacturing and innovation, 3D printing has emerged as a beacon of possibility. Its ability to turn digital designs into tangible objects has revolutionized industries from aerospace and automotive to healthcare and fashion. Yet, amid the awe-inspiring potential of this technology, a complex tapestry of challenges has woven itself into the fabric of 3D printing's journey. As we embark on a comprehensive exploration of this remarkable realm, this book seeks to unravel the intricacies of these challenges and illuminate a solution-driven path forward. At the heart of this narrative lies a pivotal concept: multi-physics optimization. As the foundation upon which the future of 3D printing is built, the importance of multi-physics optimization cannot be overstated. This book presents the world of 3D printing, dissecting its various facets and intricacies, while emphasizing the crucial role that optimization plays. Through a prism of multidisciplinary perspectives, we will explore the labyrinthine world of material properties, structural integrity, printing speed, and cost-effectiveness.

In this book, efforts have been made to navigate the nuances of characterization and optimization, uncovering the interplay of variables that define the 3D printing process. From the composition of materials to the geometry of design, every decision reverberates through the layers of this additive manufacturing journey. The book has been conceived and written to encompass all aspects of the 3D printing process – from its applications and challenges to the importance of process optimization – and is thus divided into two separate parts. The book has been conceived and written to encompass all aspects within the scope of process optimization. It consists of fourteen chapters, each focusing on core research areas grounded in current understanding and evidence. The contributions to these chapters come from key opinion leaders and international experts in the field of 3D printing, as well as industrial researchers, making this book a truly interdisciplinary collaboration. The upcoming chapters will immerse readers in strategies and techniques that illuminate the path to excellence. The book expresses the importance of multi-physics optimization, where the pursuit of a single ideal gives way to the quest for harmony among conflicting objectives. Through this lens, we will bridge the gap between theory and practice, translating academic insights into real-world applications. From intricate mathematical models to practical case studies, this book serves as a roadmap for practitioners, researchers, and enthusiasts alike.

The topics have been selected based on existing research evidence, current challenges, optimization possibilities, and future needs. The first chapter serves as an overview of the entire book, emphasizing the significance of approaching the 3D printing process as a multidisciplinary field. It discusses landmark events and contributions by pioneers in the field in an engaging and readable manner. Moving forward, the second chapter highlights the potential of transitioning 3D printing from laboratory settings to industrial applications. One of the key factors is gaining a comprehensive understanding of the various applications of 3D printing and the corresponding applicable materials. Chapters 3 and 4 delve into a review of the existing literature in this context. Chapter 5 provides an in-depth analysis of the current challenges in 3D printing. Following this, Chapter 6 elucidates the importance of multi-objective evaluation in the 3D printing process. Notably, process variables emerge as pivotal features in 3D printing, and Chapter 7 explores their role in optimization. The succeeding chapters, namely, 8, 9, 10, and 11, focus on different aspects of 3D-printed parts. Chapter 8 delves into the physico-chemical attributes of 3D-printed components, while Chapter 9 underscores the significance of rheological behavior. Chapter 10 examines how the mechanical behavior of 3D-printed parts serves as an indicator of final part quality. Chapter 11 reviews thermal behavior and the importance of heat transfer within the 3D printing process. Chapter 12 addresses the criticality of in-process temperature monitoring in assessing the quality of 3D-printed parts during layer deposition. In alignment with the book's objectives, Chapter 13 explains the importance of optimizing control factors and properties of 3D-printed components. The final chapter sheds light on the role of machine learning and its impact on optimizing the 3D printing process.

This book is envisioned as a primary and comprehensive reference source for material scientists, graduate engineers, postgraduates, research scholars, industrial engineers, and technicians working in the field of additive manufacturing and 3D printing. We have made sincere efforts to review and compile all relevant information based on currently available evidence. We hope that this book will be extremely useful for students, researchers, and industrialists engaged in the multidisciplinary process of 3D printing.

1

3D Printing as a Multidisciplinary Field

Hamid Reza Vanaei[1,2,3], *Sofiane Khelladi*[3], *and Abbas Tcharkhtchi*[4]

[1] *ESILV, Léonard de Vinci Pôle Universitaire, 92916 Paris La Défense, France*
[2] *Léonard de Vinci Pôle Universitaire Research Center, 92916 Paris La Défense, France*
[3] *Arts et Métiers Institute of Technology, CNAM, LIFSE, HESAM University, 75013 Paris, France*
[4] *Arts et Métiers Institute of Technology, CNRS, CNAM, PIMM, HESAM University, 75013 Paris, France*

1.1 Introduction

The challenges related to effective bonding, diminished strength, and mechanical performance of 3D models created through fused filament fabrication (FFF) remain significant focal points within the realm of 3D-printed structures. FFF, commonly known as 3D printing, is extensively utilized in crafting prototypes for various industries such as aerospace, medical, and automotive sectors. This technique involves feeding a thermoplastic polymer into a liquefier, which then extrudes a filament while traversing successive X–Y planes along the Z-axis direction. This process results in the gradual construction of a 3D component layer by layer. As the deposition advances, the heated filament is placed upon previously deposited filaments that are in the cooling phase. This action triggers their reheating, establishing a specific period during which the interfaces of contacting filaments achieve temperatures beyond the glass transition temperature (T_g) for amorphous materials or the crystallization temperature (T_c) for semicrystalline materials. This temperature range is essential for effective bonding to occur. Consequently, each filament must attain a sufficient level of heat during deposition, while avoiding excessive heat that might lead to deformation due to the force of gravity and the weight of subsequently deposited filaments in subsequent layers.

Hence, numerous factors influence the quality of the produced component, such as the polymer's temperature profile, and, consequently,

Industrial Strategies and Solutions for 3D Printing: Applications and Optimization, First Edition. Edited by Hamid Reza Vanaei, Sofiane Khelladi, and Abbas Tcharkhtchi.

the bonding between layers. Consequently, comprehending the manner in which the process parameters impact the temperature changes in the filaments is of significant importance. Furthermore, given that rheological properties like viscosity are influenced by temperature variations, it is plausible to establish a connection between this dependency and the changing temperatures of the deposited filaments. This suggests the possibility of establishing a linked correlation between viscosity and temperature. Consequently, this concept gives rise to the notion of a time, temperature, transformation (TTT) diagram for the material, enabling the simultaneous exploration of both temperature variations and their impact on viscosity.

While 3D printing offers certain benefits, there is a need for enhancement and refinement to meet industry standards. This improvement entails enhancing mechanical attributes and the quality of bonding (with the goal of achieving superior component quality), as well as minimizing production costs and construction duration (with the aim of optimizing the overall process).

Given the abovementioned statements, the temperature evolution during 3D printing process thoroughly specified the quality and mechanical strength of fabricated structures. Experimental monitoring and analytical investigations are still challenging in 3D printing, and lack of practical knowledge corresponds to the problem of bonding in this process. Since the rheological characteristics are a function of temperature, together with the mentioned process variables, are widely affected by temperature evolution of filaments while printing. To sum up, investigation on temperature and temperature dependence viscosity of 3D printing materials while printing is still in its early stages, and it governs the bonding quality itself.

This chapter highlights the significance of 3D printing as a multidisciplinary field. It emphasizes the pivotal variables that exert control over the bonding quality of the end product. Additionally, the discussion delves into essential techniques for characterizing these factors and enhancing our perspective for the purpose of optimization through a combination of experimental and numerical methods.

1.2 Unveiling the Foundations: Grasping the Essential Features of 3D Printing

1.2.1 Historical Review

ME-3DP emerged as one of the earliest innovations in the realm of 3D printing, with its inception traced back to the late 1980s through the appearance of its initial patent [1]. This pivotal development paved the way for the establishment of Stratasys, an entity that has since grown to attain a prominent status as one of the globe's largest and most influential enterprises in the 3D printing domain. Throughout the initial 20 years of ME-3DP's existence, Stratasys maintained a notable presence, primarily wielding its fused deposition modeling (FDM®) apparatus. These integrated machines, recognized for their association with rapid prototyping, paralleled the functions of other contemporary 3D printing technologies during that era.

A significant transformation in both the technological and market landscapes of 3D printing emerged during the later part of the 2000s. This transformative shift was primarily instigated by two pivotal occurrences – first, the advent of the RepRap initiative, which championed open-source principles, and second, the expiration of Stratasys' original patents. This confluence of events heralded a proliferation of participants in the market, primarily consisting of hardware startups, who embarked on the development of desktop 3D printers inspired by the RepRap model and bearing striking resemblances to Stratasys' machinery [2]. This period witnessed the rapid expansion of several of these entities, either propelling them to the forefront of the market or leading them to be acquired by established industry leaders. The consequence was the emergence of a burgeoning market characterized by an ever-diversifying array of products. For instance, the current landscape features an extensive array of FFF (increasingly standardized as opposed to FDM) printers, spanning a price spectrum from modest hundreds of dollars to significant figures in the hundreds of thousands. Another pivotal transformation was the transition from self-contained product architectures, emblematic of the early era of 3D printing firms, to an open ecosystem comprising specialized providers of machinery, materials, components, and software. This shift played a pivotal role in accelerating the democratization and innovation within the 3D printing field. It facilitated the technology's penetration into markets that had hitherto remained inaccessible due to various barriers. Presently, 3D printing stands as the most widely embraced 3D printing technology, quantified by both the staggering volume of machine shipments (exceeding one million annually on a global scale) and the extensive user base numbering in the millions.

For the 3D printers, a considerable proportion is comprised of FFF printers that, for the most part, retain operational similarities to the original Stratasys machines. Nevertheless, there have been successive innovations within the process over time, resulting in a significant broadening and enhancement of the technological landscape governing 3D printing methodologies (as indicated in Figure 1.1). Among the most prominent

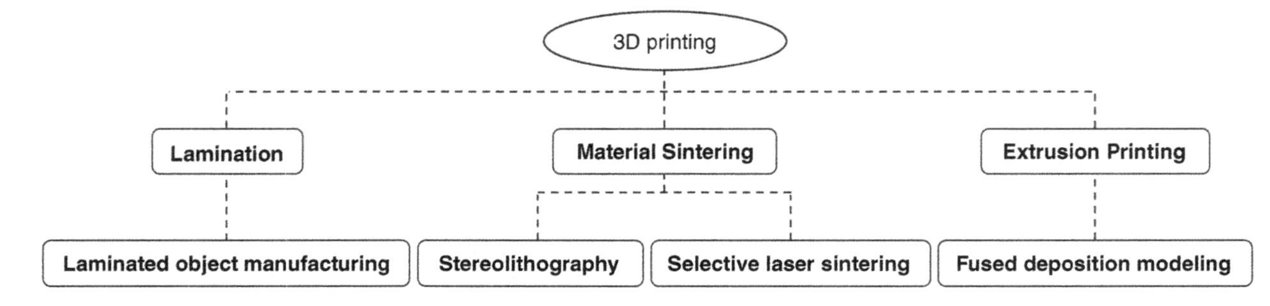

Figure 1.1 Process variations in 3D printing.

instances is the utilization of 3D printing for generating continuous fiber-reinforced composite structures, a groundbreaking concept initially pioneered by the US-based firm Markforged. Subsequently, this innovation garnered widespread attention both within industrial circles and scholarly discourse. This advancement empowers the creation of structures characterized by extraordinary stiffness and resilience, thereby enabling 3D printing to venture into novel domains of material attributes that had hitherto remained inaccessible.

1.2.2 Potential of 3D Printing from Lab to Industry

The evolution of 3D printing technology, from its origins in laboratories to its widespread adoption in various industries, has been nothing short of transformative. Initially developed as a rapid prototyping tool in the 1980s, 3D printing has now emerged as a powerful force that holds immense potential across a wide spectrum of applications. Its journey from lab experimentation to industrial integration highlights its capacity to revolutionize traditional manufacturing processes and redefine the boundaries of creativity and innovation.

In the early stages, 3D printing was primarily confined to research laboratories, where it was used for creating prototypes and conceptual models with relative ease and speed. This capability drastically cuts down development times and costs compared to traditional subtractive manufacturing methods. As the technology advanced, its potential began to extend beyond prototyping, leading to the production of intricate and customized components that were previously deemed unfeasible. As 3D printing techniques evolved, industries such as aerospace, automotive, healthcare, and consumer goods took notice of the technology's promise. In aerospace, for instance, the ability to create lightweight, complex geometries with high precision has enabled the production of fuel-efficient components and reduced overall aircraft weight. In healthcare, 3D printing has enabled the creation of patient-specific implants, prosthetics, and medical models for surgical planning. This personalized approach has improved patient outcomes and streamlined medical procedures.

One of the most significant advantages of 3D printing is its potential to revolutionize supply chains and manufacturing processes. Traditional manufacturing often involves producing parts in large quantities and shipping them globally, which can be time-consuming and costly. With 3D printing, companies can transition to localized manufacturing, producing items on demand and reducing the need for large inventories and long shipping distances. This not only reduces waste but also allows for rapid response to market demands and customization. Furthermore, 3D printing opens doors to intricate and innovative designs that were previously constrained by the limitations of traditional manufacturing methods. Complex structures, lattice designs, and organic shapes that were once impossible or economically unviable to create can now be realized with 3D printing. This encourages designers to push the boundaries of creativity, resulting in products that are not only functional but also aesthetically captivating. However, while the potential of 3D printing is vast, challenges still exist. Material limitations, production speed, and post-processing requirements are areas that researchers and industries are actively addressing to unlock even greater potential. As materials improve and new techniques emerge, 3D printing's role is likely to expand, influencing industries beyond manufacturing, such as architecture, fashion, and even food. The journey of 3D printing from laboratory experimentation to industrial integration showcases its potential to disrupt traditional manufacturing, accelerate innovation, and offer new levels of customization. Its ability to produce complex, personalized, and functional objects has paved the way for advancements across diverse sectors. As research and development continue, the full extent of 3D printing's impact on industries and society at large is only beginning to unfold.

1.2.3 Challenges and Potential Roadmap Toward Solving them in 3D Printing

The rapid advancement of 3D printing technology has undeniably brought about transformative changes across various industries. However, this progress has not been without its share of challenges, many of which have become focal points for researchers, engineers, and industry professionals aiming to further harness the potential of additive manufacturing.

I. **Material Limitations:** The range of materials available for 3D printing is expanding, but it remains limited compared to traditional manufacturing processes. While some materials like plastics and metals are well established, there is a demand for more diverse and specialized materials that can meet the requirements of specific applications, such as high-temperature environments, biocompatibility, or electrical conductivity.

II. **Print Speed and Scale:** 3D printing is often criticized for its relatively slow production speeds, particularly when compared to traditional manufacturing methods. The layer-by-layer additive process inherently takes longer, making it less suitable for high-volume, time-sensitive production. Scaling up 3D printing while maintaining quality and efficiency remains a challenge.

III. **Surface Finish and Post-Processing:** Parts produced by 3D printing may exhibit rough surfaces and layer lines. Achieving a smooth and high-quality surface finish without extensive post-processing is a challenge. Additional steps like sanding, polishing, or chemical treatments are often required, increasing the overall production time and complexity.

IV. **Structural Integrity and Quality Control:** Ensuring the structural integrity and consistency of 3D-printed parts is essential, especially for critical applications in industries like aerospace and healthcare. Variations in layer adhesion, porosity, and material properties can impact the mechanical performance of the final product. Developing reliable quality control measures to identify defects and inconsistencies is an ongoing concern.

V. **Design Complexity and Optimization:** While 3D printing enables intricate designs and geometries, not all designs are inherently suited for additive manufacturing. Designing for 3D printing requires a different approach, taking into account factors like overhangs, support structures, and material properties. Optimizing designs for both functionality and printability is a challenge that demands specialized skills and software tools.

VI. **Regulatory Hurdles:** Industries like healthcare and aerospace are subject to strict regulatory standards. Introducing 3D-printed components into these sectors requires demonstrating their reliability, safety, and compliance with regulatory requirements. Developing standardized testing and certification procedures for 3D-printed products is an ongoing endeavor.

VII. **Cost Considerations:** While 3D printing can be cost-effective for low-volume production and customized items, it might not always be the most economical option for mass production. The costs associated with materials, equipment, and post-processing can be significant, requiring a thorough analysis of cost-effectiveness for each application.

VIII. **Sustainability and Material Recycling:** The environmental impact of 3D printing, particularly in terms of material waste and energy consumption, is a concern. Many 3D printing processes generate excess material that cannot be reused, leading to material waste. Developing more sustainable materials and recycling methods is crucial to mitigate this challenge.

The journey toward confronting the fundamental challenges intrinsic to the 3D printing process has catalyzed the formulation of potential pathways, each charting a direction to unleash the technology's complete potential. To surmount the obstacles associated with sluggish construction rates and properties displaying anisotropy, a strategic trajectory involves delving into the realm of advanced materials engineering. The exploration of innovative materials characterized by heightened ease of printing and optimized thermal attributes holds the potential to accelerate deposition rates while alleviating anisotropy concerns. In parallel, the enhancement of extrusion mechanisms, capitalizing on multi-nozzle systems or continuous extrusion methodologies, has the potential to overhaul the speed of material deposition and lead to the creation of components possessing greater isotropy. Running in tandem with advancements in materials, the course to address inconsistencies in mechanical properties entails the cultivation of robust interlayer bonding techniques. Research endeavors are honing in on the comprehension of molecular diffusion at interfaces, the optimization of thermal conditions, and the design of sophisticated extruders engineered to foster more robust adhesion between layers. These collective endeavors, coupled with the integration of predictive modeling and simulation, seek to cultivate a deeper understanding of the intricate interplay linking process parameters and mechanical attributes. Ultimately, these pursuits aspire to engender heightened reliability and predictability, culminating in more assured and foreseeable outcomes.

To overcome the obstacle of uncertainty surrounding part characteristics, a strategic approach directs attention toward innovation driven by data. Leveraging the capabilities of machine learning (ML) and artificial intelligence (AI), scientists are engaged in efforts to establish connections among process factors, material attributes, and eventual part behavior. By assimilating extensive datasets, these techniques hold the potential to unveil patterns and revelations, facilitating the development of resilient models capable of predicting part attributes based on distinct printing parameters.

As the landscape of 3D printing undergoes transformation, another strategic direction envisions a convergence involving hardware, software, and process regulation. Progressions in real-time monitoring systems, adaptable control algorithms, and closed-loop feedback mechanisms stand ready to augment the precision and uniformity of printing. These advancements possess the potential to effectively counterbalance discrepancies stemming from factors such as fluctuations in temperature or inconsistencies in materials, thereby fostering the production of components marked by heightened accuracy and dependability. Essentially, these potential pathways embody collaborative endeavors to navigate the intricate realm of 3D printing challenges. Through the harmonization of inventive materials, sophisticated interlayer bonding techniques, insights derived from data analysis, and refined process management, the envisioned path ahead anticipates a future in which 3D printing not only surmounts its hurdles but also catalyzes industries with unparalleled swiftness, precision, and adaptability.

1.2.3.1 High Building Rate 3D Printing Process

The building rate in 3D printing, a crucial factor determining the speed of the additive manufacturing process, is inherently constrained by the capabilities of the printing system's module. In this context, enhancing the performance of individual modules within the system holds the potential to yield substantial improvements in the overall building rate of 3D printing [3, 4]. This acknowledgment underscores the significance of focusing on module-level advancements to expedite the additive manufacturing process.

Presently, two promising avenues emerge as potential technologies capable of realizing higher building rates within the 3D printing process. These avenues represent distinct approaches that could potentially revolutionize the speed and efficiency of the technology. By addressing the challenges inherent to low building rates, these technologies aim to enhance the competitiveness of 3D printing as a viable manufacturing method across various industries. As the pursuit of faster additive manufacturing continues, these innovative pathways promise to reshape the landscape of 3D printing, potentially catapulting the technology into new realms of application and significance.

1.2.3.2 Big Area Additive Manufacturing (BAAM) System

Big Area Additive Manufacturing (BAAM) stands out as a notable advancement in the realm of large-format printing systems, pioneered by Oak Ridge National Laboratory. Designed to produce structures on a scale of several meters, this technology introduces the capability to swiftly print sizable components while maintaining high extrusion rates [5]. The unique approach of utilizing feedstock in the form of thermoplastic or reinforced thermoplastic pellets brings about a substantial reduction in material deposition costs, presenting a cost-efficient alternative that slashes expenses by a factor of 20 times. At the core of this system lies a single-screw extruder, responsible for both melting the plastic pellets and subsequently extruding the molten material onto a heated print bed. This single-screw extruder further distinguishes itself by its remarkable deposition speed, surpassing conventional polymer additive manufacturing systems by over 200 times and boasting rates of up to 50 kg/h.

An additional feature contributing to BAAM's prowess is its expansive build platform, capable of accommodating structures of impressive dimensions. With the potential to construct objects as substantial as 6 m in length, 2.4 m in width, and 1.8 m in height, BAAM's build capacity surpasses most commercial systems by a factor of approximately 10. This expanded envelope of production capability positions BAAM as a technology with promising implications across diverse sectors, including automotive, aerospace, and energy industries. The exceptional attributes of BAAM, from its rapid extrusion rates and economical material usage to its capacity for fabricating sizable structures, underscore its potential to revolutionize manufacturing methodologies, making substantial contributions to the realms of transportation, aeronautics, and power generation.

1.2.3.3 Faster FFF 3D Printing System

To surmount the constraints imposed by performance-limiting mechanisms, the development of a swift FFF system hinges on several key enhancements. These enhancements encompass an improved material feed mechanism, augmented heat transfer rates from the liquefier wall to the filament core, and the optimization of gantry positioning systems guided by open-loop stepper motors. Researchers' estimations suggest that a remarkable 10-fold increase in building rate compared to typical commercial desktop systems could be attainable through the enhancement of printer module performance and the implementation of refined toolpath planning algorithms [5]. In parallel, innovative hardware components have emerged to bolster the potential of swift FFF systems. Extruders featuring multipoint contact with the filament have been introduced to the market, ensuring a firm grip on the filament, and preventing slippage during the printing process. Furthermore, the liquefier, now equipped with dual heat zones and meticulous temperature control mechanisms, maintains the polymer at a consistent temperature throughout the extrusion process. The incorporation of Step-Motor Driving Systems aids in achieving elevated velocity and acceleration for the stepper motor, ultimately allowing for printing speeds reaching up to 250 mm/s.

Recently, a rapid FFF system boasting a building rate of approximately 720 cm^3/h has become available on the market. Looking ahead, it is foreseeable that further advancements in module functionality will pave the way for even swifter FFF systems. The integration of AI-driven assistance systems, such as Camera Recognition, in conjunction with refined path planning algorithms, holds promise for significantly elevating building rates. As these technological advancements continue to converge, it is reasonable to anticipate the emergence of fast FFF systems with progressively higher building rates, propelling the field of additive manufacturing into a new era of enhanced speed and efficiency.

1.2.3.4 Improvement of Interfacial Bonding and Strength in Z-Direction

Examining the underlying factors influencing the weakened bond interface in the context of ME-3DP underscores the need for a dual-pronged approach encompassing both process control and judicious material selection. To effectively bolster the interfacial bonding strength and concurrently diminish the pronounced anisotropy characteristic of printed components, these two key aspects must be thoughtfully addressed. Process control plays a pivotal role in sculpting the final quality of ME-3D printed parts. Fine-tuning various printing parameters, such as extrusion temperature, layer height, and print speed, can exert a significant influence on the formation of robust interlayer bonds. By meticulously optimizing these parameters, the thermal conditions during printing can be adjusted to facilitate enhanced molecular chain diffusion at the interface, leading to stronger and more dependable bonds between successive layers. Process control also extends to real-time monitoring and feedback mechanisms, allowing for dynamic adjustments during printing to counteract potential deviations and ensure consistent interfacial bonding throughout the entire production process.

Simultaneously, the choice of materials assumes paramount importance in augmenting interfacial bonding strength. The selection of polymers with compatible melt viscosities and thermal properties can foster more cohesive bonding between layers. Additionally, the incorporation of additives, such as coupling agents or compatibilizers, can promote molecular adhesion and alignment, further enhancing the overall interlayer adhesion. Furthermore, materials with balanced mechanical properties and thermal behaviors across multiple directions can contribute to reducing anisotropy, resulting in more uniform and predictable mechanical performance across all orientations.

Harmonizing process control and material selection leverages a comprehensive strategy to surmount the challenge of weak bond interfaces and the ensuing anisotropy in ME-3D printed parts. This combined effort, founded on precise control over the printing process parameters and the strategic utilization of materials with optimal properties, offers a pathway toward producing components with heightened strength, reliability, and uniformity across diverse orientations.

The inadequate adhesion between layers stems from the limited diffusion of large and sluggish polymer chains across the interface between filaments. Enhancing diffusion and interlayer adhesion in FDM-fabricated parts can be facilitated by employing filament materials that contain lower molecular weight components. This choice is driven by the fact that polymers with lower molecular weights exhibit quicker diffusion across the inter-filament interface, thereby fostering improved interlayer adhesion [6]. A promising solution to address the weakness in the z-direction involves utilizing filaments with a core-shell structure, a concept realized through polymer coextrusion [7–10]. Typically, the filament's core possesses a notably higher glass transition temperature than its sheath, imparting favorable attributes such as enhanced printability, stiffness, and dimensional stability. The interplay between these dual materials benefits from well-established chemical compatibility, often resulting in a phase-separated blend characterized by excellent adhesion between phases. Predominantly, the most prevalent combination involves a dual-material filament comprising ABS with a PC core. This combination is frequently subjected to annealing post-printing, a treatment that reinforces bonding between printed layers while upholding dimensional stability.

Compared to conventionally printed single-material filaments, components produced using the dual-material filament exhibit remarkable enhancements. For instance, post-printing annealing culminates in a fivefold amplification in z-direction impact toughness and a fourfold increase in z-direction tensile strength. Impressively, these benefits are achieved without compromising part dimensional accuracy or surface quality. Notably, even when elevated printer nozzle temperatures are employed, the inherent stability of the core material exerts a stabilizing effect, ensuring excellent dimensional precision and surface finish.

1.2.4 Role of Controlling Factors in 3D Printing

In the domain of 3D printing, the influence of controlling factors takes on paramount significance, exerting a profound and undeniable impact on the very essence of the additive manufacturing process. This influence spans across an intricate web of parameters, encompassing the diverse domains of material properties, design intricacies, printing configurations, and post-processing methodologies. This intricate interplay among these varied factors is the architect of the final outcomes, ultimately shaping the characteristics, quality, functionality, and efficiency of the fabricated objects. In the relentless pursuit of crafting components that not only meet but surpass performance expectations, the optimization of these controlling factors emerges as a central tenet. Material properties stand as a foundational cornerstone, wherein the careful selection of materials becomes the maestro that dictates the mechanical traits, thermal dynamics, and the flow behavior of materials during the deposition process. Concurrently, design parameters weave a narrative of both aesthetics and functionality, wielding the power to define internal structures, intricate infill patterns, and overarching geometrical configurations.

The strategic manipulation of printing parameters constitutes another layer of finesse. The orchestration of variables such as layer thickness, print speed, and temperature acts as a symphony conductor, fine-tuning the deposition process to orchestrate factors like layer adhesion, surface finish, and the mechanical robustness of the final product. However, the journey does not conclude there; it extends into the post-processing realm. Post-processing techniques provide a canvas for further enhancing properties and refining aesthetics, emphasizing the holistic nature of the optimization endeavor.

As these controlling factors converge and collaborate, a harmonious symphony emerges. This orchestration encapsulates the essence of additive manufacturing, empowering the creation of parts that bear meticulously tailored attributes. This harmonious synthesis resonates perfectly with the overarching ambition of 3D printing itself – the art of meticulously crafting components endowed with impeccable precision, bespoke designs, and elevated performance characteristics. In the continued evolution of this field, the comprehension and mastery of these controlling factors will persist as a central fulcrum, opening pathways for innovation across a wide spectrum of industries. In doing so, it ushers in a new era, where the realm of advanced manufacturing thrives with limitless possibilities.

1.3 Multiphysics Behavior in 3D Printing Process

1.3.1 Physicochemical and Mechanical Phenomena of 3D-printed Parts

Physicochemical and mechanical phenomena of 3D-printed parts are pivotal factors that define the overall quality and functionality of the additive manufacturing process. These phenomena encompass a multidimensional spectrum of material behavior and structural attributes, fundamentally shaping the performance and application potential of the final components. Physicochemical attributes delve into the intricate properties of the materials used in 3D printing. Factors such as material composition, thermal properties, and molecular structure play a vital role in dictating how the material responds during the printing process. Understanding these phenomena is essential for achieving optimal print quality, as they influence aspects like material flow, adhesion to previous layers, and susceptibility to heat-related issues like warping. Mechanical characteristics, on the other hand, are the mechanical properties that the printed parts exhibit once they are completed. This encompasses a diverse range of attributes including tensile strength, hardness, flexibility, and resilience. These mechanical traits are closely tied to the material properties and how they interact with the printing process itself. Achieving the desired mechanical properties often involves striking a delicate balance between factors like print temperature, layer adhesion, infill density, and post-processing techniques.

In the quest for robust and reliable 3D-printed parts, a comprehensive understanding of both the physicochemical and mechanical phenomena is paramount. Engineers and researchers work meticulously to optimize these attributes to meet specific performance requirements for various applications. Tailoring material choices, design considerations, and printing parameters to align with the intended function of the part is an ongoing pursuit to ensure that the 3D-printed components not only achieve the desired mechanical robustness but also exhibit the necessary physicochemical attributes to withstand real-world conditions and demands. As 3D printing technology continues to advance, the ability to control and manipulate these physicochemical and mechanical phenomena becomes increasingly refined. This progress opens doors to a myriad of applications across industries, ranging from aerospace to healthcare and beyond, where the tailored properties of 3D-printed parts enable innovations that were previously unattainable through traditional manufacturing methods.

1.3.2 Thermal Features of 3D-printed Parts

3D printing process involves the utilization of a thermoplastic polymer system, often in the form of circular filaments, pellets, or powder. These polymer materials undergo a sequence of transformation steps – they are compressed, heated until molten, and then extruded as a filament thread through a nozzle. Concurrently, a build platform, known as the latter, is directed along a predetermined printing trajectory. This orchestrated movement forms the foundation upon which three-dimensional structures come into existence. These structures are constructed as a sequence of vertical layers, each layer comprising horizontally arranged cross sections composed of the deposited threads. This method permits the fabrication of geometrically intricate components, showcasing the power of this manufacturing technique.

Central to the integrity of the resulting parts is the bonding that takes place between the adjacent threads. This inter-thread bonding is essential for the physical coherence of the printed part. The mechanical properties, surface texture, dimensional precision, and residual stresses inherent in a 3D-printed object are notably impacted by the quality of bonding achieved and the specific morphology that evolves during the cooling phase. To establish effective bonding, an extensive process of diffusion and entanglement of polymer macromolecules transpires across the interfaces of neighboring filament threads. This intricate process relies on a combination of factors, including the rheological characteristics of the polymer material, the local temperatures experienced, and the associated timeframes. In essence, for a given part geometry to be printed within the confines of specific processing conditions, 3D printing encapsulates a series of interconnected stages. This involves the sequential phases of material heating and melting, the application of melt pressure as it flows through a nozzle, the deposition of the extruded thread onto a build platform, and the subsequent layering upon previously deposited threads. Cooling and solidification mark the culmination of this multifaceted journey.

The focus for thermal characterization requires to be on the comprehensive modeling of the thermal dynamics encompassing the deposition and cooling stages within the 3D printing process. The overarching goal is to unravel the intricate correlations interwoven among printing parameters, the resultant temperature profiles, and the extent of bonding achieved. Such thermal modeling holds the potential to significantly enhance our comprehension of the 3D printing process, enabling better-informed decisions regarding part design, the selection of corresponding printing parameters, and optimizing the manufacturing process itself. By unraveling the thermal intricacies, this approach opens pathways to more precisely tailored parts and more efficient manufacturing processes within the domain of material extrusion additive manufacturing.

1.3.3 Rheological Evaluations in 3D Printing

Rheological characteristics form the cornerstone of successful 3D printing, governing how materials flow, deform, and interact during the additive manufacturing process. In the realm of 3D printing, where intricate structures are built layer by layer, understanding the rheological behavior of the materials used is paramount. This page delves into the significance of rheological properties in 3D printing, exploring key concepts such as viscosity, shear thinning, and thixotropy.

1.3.3.1 Mastering the Flow: Essential Fundamentals of Rheology

Viscosity is considered a foundational rheological property and dictates a material's resistance to flow. In the context of 3D printing, viscosity directly influences the extrusion process. The optimal viscosity range is crucial for ensuring smooth material flow through the printing nozzle. A material that is too viscous might lead to inconsistent extrusion, clogs, or insufficient layer bonding. Conversely, a material with low viscosity might result in excessive spreading and poor adhesion between layers. Achieving the right viscosity balance is vital for maintaining the dimensional accuracy and surface finish of printed parts. Besides, shear thinning is a prevalent rheological behavior in 3D printing materials and offers a unique advantage. It refers to the phenomenon where a material's viscosity decreases as the shear rate increases. This property facilitates seamless extrusion during high-speed nozzle movements, allowing for precise deposition of material. When the extrusion halts, the material's viscosity rebounds, preventing unwanted material flow. Shear thinning contributes to the accurate reproduction of intricate details, resulting in parts with enhanced geometric fidelity, structural integrity, and surface finish.

1.3.3.2 Optimizing with Rheological Insights

It is crucial to consider a deep understanding of rheological characteristics as it empowers manufacturers to optimize their 3D printing processes and achieve superior results. Thixotropy, an intriguing rheological attribute, refers to a material's ability to exhibit decreased viscosity over time under constant shear stress. In the context of 3D printing, thixotropy plays a role in material recovery after extrusion pauses. This property prevents drooping or sagging of material when the extruder is stationary, preserving the structural integrity of the printed layers. Thixotropic materials ensure stability during printing pauses and contribute to the overall success of intricate designs.

To harness the full potential of rheological characteristics, manufacturers rely on rheological testing methods. These techniques quantify how materials respond to temperature changes, shear rates, and stress. By generating comprehensive rheological profiles, manufacturers can make informed decisions about printing parameters, nozzle sizes, and even material formulations. Additionally, advanced modeling and simulation tools predict how materials will behave during the printing process, allowing for virtual experimentation and optimization of printing strategies.

In the dynamic domain of 3D printing, a profound comprehension of rheological characteristics is indispensable. Viscosity, shear thinning, and thixotropy collectively shape the material's behavior, directly influencing print quality, precision, and mechanical properties. By leveraging these insights, manufacturers can tailor their printing strategies to suit a material's rheological behavior, achieving exceptional print outcomes across diverse applications. Rheological mastery represents a bridge between scientific understanding and the art of additive manufacturing, paving the way for groundbreaking innovations in the world of 3D printing.

1.3.4 In-process Temperature Monitoring in 3D Printing

In-process temperature monitoring is a crucial aspect of enhancing the precision, quality, and reliability of 3D printing processes. 3D printing, also known as additive manufacturing, involves the layer-by-layer deposition of materials to create intricate three-dimensional objects. The temperature at various stages of the printing process significantly impacts the structural integrity, dimensional accuracy, and material properties of the final printed object. As a result, real-time temperature monitoring and control have emerged as fundamental techniques to ensure consistent and optimal printing outcomes. During 3D printing, the temperature of the print bed, the extruder or nozzle, and the ambient environment all play pivotal roles in determining the final product's characteristics. In-process temperature monitoring involves the use of sensors, such as thermocouples or infrared cameras, to capture and analyze temperature data at critical locations. These sensors provide real-time feedback to the printer's control system, enabling it to make necessary adjustments to temperature settings on the fly. This continuous monitoring helps prevent issues like warping, layer misalignment, or insufficient material bonding, which can arise due to temperature fluctuations.

One of the primary advantages of in-process temperature monitoring is its ability to ensure consistent material properties throughout the printed object. Different materials used in 3D printing, such as plastics, metals, and composites, exhibit specific thermal behaviors that influence their flow characteristics and solidification rates. By closely monitoring and controlling temperatures, operators can mitigate material-specific challenges, such as nozzle clogging or uneven cooling, leading to more predictable and repeatable printing outcomes. Furthermore, in-process temperature monitoring enables the identification of anomalies or deviations from desired temperature profiles. If a temperature excursion occurs, the monitoring system can trigger alerts or pause the printing process, preventing the production of defective parts. This capability is especially important in industrial applications where quality control is paramount and production efficiency is closely tied to minimized waste and downtime.

Presumably, in-process temperature monitoring is a critical advancement in 3D printing technology. By maintaining precise temperature conditions throughout the printing process, manufacturers can achieve higher-quality prints, reduce waste, and enhance the overall efficiency of additive manufacturing. As 3D printing continues to evolve and find applications in various industries, the role of in-process temperature monitoring is set to become even more integral in ensuring the reliability and consistency of printed products.

1.4 3D Printing Perfection: Unveiling the Power of Optimization

1.4.1 Importance of Multiphysics Evaluation in 3D Printing

In 3D printing, multiphysics evaluation emerges as a crucial cornerstone for informed decision-making and optimized outcomes. Unlike traditional manufacturing processes, 3D printing offers a spectrum of possibilities across various dimensions, ranging from material choices and part geometries to production speeds and cost considerations. Embracing a multiphysics approach acknowledges the intricate interplay between these diverse factors and strives to strike a delicate balance that meets a multitude of criteria. One of the most significant advantages of multiphysics evaluation lies in its ability to navigate the complex web of trade-offs. Inherent in 3D printing are numerous competing physics, such as achieving high material strength while minimizing weight or enhancing production speed while maintaining impeccable surface finish. By adopting a multiphysics lens, manufacturers and designers can gain a comprehensive understanding of how adjustments in one area affect others, leading to more informed compromises that align with overall goals.

This approach also allows for a holistic examination of the 3D printing process, from design to post-processing. For instance, in aerospace applications, a multiphysics evaluation can consider factors like weight reduction, thermal resistance, and material compatibility. By assessing these multiple dimensions simultaneously, stakeholders can refine designs that not only meet stringent performance requirements but also optimize manufacturing efficiency and economic viability. Moreover, in industries focused on sustainable practices, multiphysics evaluation becomes a vital tool. Environmental impact, material recyclability, and energy efficiency can all be weighed alongside traditional performance metrics. This enables the creation of 3D-printed products that are not only high-performing but also aligned with eco-friendly principles, a crucial aspect in a world increasingly concerned with minimizing its carbon footprint. Furthermore, multiphysics evaluation fosters collaboration and communication among interdisciplinary teams. Engineers, designers, material scientists, and business strategists can collectively analyze various aspects of 3D printing projects, enabling a more comprehensive understanding of potential challenges and opportunities. This collaborative effort enhances creativity, reduces blind spots, and often leads to innovative solutions that may have otherwise been overlooked.

As the 3D printing landscape continues to evolve, adopting a multiphysics evaluation approach becomes increasingly imperative. It empowers stakeholders to make informed decisions that account for the multifaceted nature of the technology and its applications. Whether in industries focused on cutting-edge innovation, sustainability, or cost-effective manufacturing, embracing multiphysics evaluation paves the way for a more comprehensive, strategic, and successful journey into the realm of 3D printing. In summary, following criteria could be taken into account to include all the variables that are engaged in 3D printing process:

1. **Balancing Diverse Physics:** Multiphysics evaluation in 3D printing enables the balanced consideration of various objectives, accommodating the intricate trade-offs between factors like cost, quality, and sustainability.
2. **Holistic Decision-Making:** By integrating multiple criteria, multiphysics evaluation empowers a holistic approach to decision-making in 3D printing, ensuring a comprehensive understanding of the broader implications of choices.
3. **Optimizing Performance:** Multiphysics evaluation enhances the capability to optimize performance across different dimensions, such as material properties, structural integrity, and production efficiency.
4. **Flexibility and Customization**: Embracing a multiphysics perspective allows for the flexibility to tailor 3D printing processes to meet specific requirements, resulting in customized solutions that align with diverse needs.
5. **Informed Trade-offs:** Multiphysics evaluation provides insights into the trade-offs involved when making decisions in 3D printing, and guiding stakeholders in making informed choices that align with strategic goals.
6. **Risk Mitigation:** Incorporating multiple physics aids in mitigating risks by identifying potential vulnerabilities across various aspects of 3D printing processes and products.
7. **Sustainable Innovation:** Multiphysics evaluation fosters sustainable innovation in 3D printing, facilitating the development of solutions that balance economic, environmental, and social factors.

8. **Enhancing Collaboration:** Multiphysics evaluation encourages collaboration among interdisciplinary teams, fostering a more comprehensive understanding of the intricacies involved in 3D printing projects.
9. **Meeting Diverse Market Needs:** The multiphysics approach ensures that 3D printing solutions cater to a wide range of market demands, addressing varying customer preferences and industry requirements.
10. **Future-Proofing Strategies:** Embracing multiphysics evaluation safeguards against potential challenges and uncertainties, allowing 3D printing strategies to adapt and evolve in a rapidly changing landscape.

1.4.2 Optimizing the Controlling Factors and Characteristics of 3D-printed Parts

Optimizing the controlling factors and characteristics of 3D-printed parts is a paramount pursuit in the additive manufacturing landscape. As the technology evolves, achieving the perfect synergy between various parameters becomes pivotal in unlocking the full potential of 3D printing for diverse applications. This optimization process spans a multifaceted spectrum, encompassing material properties, design considerations, printing parameters, and post-processing techniques. At the heart of this endeavor lies material selection. Different materials exhibit varying behaviors during the printing process and influence the mechanical, thermal, and aesthetic qualities of the final product. By meticulously choosing materials that align with specific application requirements, engineers can ensure that the 3D-printed parts meet the desired performance criteria. This involves evaluating factors like material strength, flexibility, thermal conductivity, and even biocompatibility for applications in fields like healthcare.

Design intricacies also hold a significant place in the optimization journey. Tailoring designs for 3D printing can lead to reduced material usage, enhanced structural integrity, and improved functionality. By employing techniques such as topology optimization, lattice structures, and generative design, designers can create parts that not only perform optimally but are also print-friendly, mitigating challenges like overhangs and excessive support structures. The tuning of printing parameters constitutes another pivotal dimension. Variables like layer height, print speed, and nozzle temperature profoundly influence the quality, accuracy, and surface finish of 3D-printed parts. Optimizing these parameters involves striking a delicate balance between factors like print time and precision. With each adjustment, engineers can fine-tune the process to achieve the desired balance between speed, quality, and material usage. Post-processing techniques must also be considered to fully optimize 3D-printed parts. Processes like sanding, polishing, and chemical treatments can enhance surface finish, while heat treatments or annealing can optimize material properties. These techniques, when tailored to specific materials and designs, contribute to achieving the desired aesthetic and functional attributes of the final part.

To comprehensively optimize the controlling factors and characteristics, a holistic approach is essential. Often, it requires an iterative process of testing, analysis, and refinement. Advanced tools like computer simulations and predictive modeling can play a vital role in this journey, providing insights into the potential outcomes of different parameter configurations before physical production even begins. The optimization of controlling factors and characteristics in 3D-printed parts embodies a multifaceted endeavor that demands a deep understanding of materials, design principles, printing techniques, and post-processing methods. By aligning these factors to meet specific objectives, engineers and designers can harness the full power of 3D printing to create components that excel in terms of performance, efficiency, and innovation.

1.4.3 Role of Machine Learning in 3D Printing

The role of ML in 3D printing has emerged as a transformative force, reshaping the landscape of additive manufacturing (AM) by bringing intelligence and adaptability to the process. ML, a subset of AI, entails the development of algorithms that enable systems to learn from data and experiences, ultimately making informed decisions and predictions. In the context of 3D printing, ML facilitates a range of advancements that enhance efficiency, precision, and innovation.

One of the primary roles of ML in 3D printing is process optimization. By analyzing large volumes of data from previous prints, ML algorithms can identify patterns and relationships between various process parameters and the quality of the final printed part. This enables manufacturers to fine-tune printing parameters, such as layer height, print speed, and temperature, to achieve optimal outcomes in terms of mechanical strength, surface finish, and dimensional accuracy. Quality control is another crucial area where ML shines. By analyzing sensor data, ML algorithms can detect anomalies or defects in real time during the printing process. This capability enables immediate adjustments, preventing the production of faulty parts and reducing waste. ML-powered quality control also offers a higher level of consistency, which is essential for industries like aerospace and healthcare that demand stringent quality standards.

ML also plays a pivotal role in design optimization. Generative design algorithms, driven by ML, can automatically generate and refine designs based on specified criteria. This leads to innovative designs that are often beyond human imagination, optimized for both performance and efficient material usage. The iterative nature of generative design, coupled with ML's ability to learn from each iteration, accelerates the innovation cycle.

Furthermore, ML is revolutionizing material development in 3D printing. By analyzing material properties, behavior, and performance, ML can assist in identifying suitable materials for specific applications. This not only accelerates material discovery but also leads to the creation of new materials with tailored properties, expanding the possibilities of what can be achieved through additive manufacturing. In essence, the integration of ML into 3D printing heralds a new era of smart manufacturing. It empowers the technology to adapt, learn, and optimize in real time, resulting in enhanced precision, reduced costs, faster innovation cycles, and higher-quality end products. As ML continues to advance, its role in shaping the future of 3D printing is poised to have far-reaching implications across industries and applications.

1.5 Future Outlook

1.5.1 Emerging Horizons in Multidisciplinary 3D Printing

As we stand on the precipice of technological innovation, the landscape of 3D printing is rapidly evolving into a multidisciplinary field that transcends the boundaries of industries and domains. The amalgamation of diverse expertise, cutting-edge research, and collaborative efforts is propelling 3D printing toward new frontiers of possibility. This page delves into the future outlook of 3D printing as a multidisciplinary endeavor, exploring the fusion of fields and the transformational potential it holds. The future of 3D printing lies in the seamless integration of various disciplines, ranging from materials science, mechanical engineering, and computer science to biology, medicine, architecture, and beyond. The intersections between these domains are giving rise to groundbreaking applications. Imagine bioengineers and material scientists codesigning biocompatible scaffolds for regenerative medicine or aerospace engineers collaborating with material experts to revolutionize lightweight yet robust components for spacecraft. The fusion of diverse knowledge pools is a fertile ground for innovation, as it brings together novel perspectives, creative problem-solving approaches, and cross-pollination of ideas.

1.5.2 Building Life with Precision

One of the most promising avenues where multidisciplinary 3D printing is taking a giant leap is bioprinting. This intricate marriage of biology, medicine, and engineering has the potential to reshape the healthcare landscape. Scientists are working toward printing functional tissues, organs, and even intricately vascularized constructs. This could herald a new era of personalized medicine, where replacement organs are tailored to an individual's unique biology, reducing transplant rejection rates, and eliminating lengthy waiting lists. As biologists, engineers, and medical professionals collaborate, the convergence of these fields is unlocking previously unimaginable possibilities in the realm of human health.

1.5.3 Architectural Revolution: Design and Construction Reimagined

The domain of architecture and construction is experiencing a paradigm shift thanks to 3D printing's multidisciplinary embrace. Architects, structural engineers, and material scientists are collaborating to create sustainable, intricate structures with reduced waste and faster construction times. Imagine entire communities built using 3D printing techniques, utilizing locally sourced materials to address housing shortages and environmental concerns. The fusion of architectural vision, engineering precision, and material innovation is paving the way for a future where the skylines are defined by creativity, efficiency, and sustainability.

1.5.4 Sustainable Manufacturing: A Green Revolution

The principles of multidisciplinary 3D printing are even infiltrating the manufacturing sector, where the convergence of materials science, mechanical engineering, and environmental studies is heralding a sustainable revolution. Additive manufacturing techniques reduce material waste by producing objects layer-by-layer, only using the exact amount of material required. Additionally, the concept of circular economy is gaining ground, with products designed for disassembly and recycling, minimizing environmental impact. This harmony of expertise is reshaping the traditional manufacturing landscape, offering eco-friendly alternatives that align with the demands of a greener future.

1.6 Summary and Outlooks: Pioneering a Multidisciplinary Renaissance

In the grand tapestry of technological progress, 3D printing's trajectory as a multidisciplinary field is a testament to human ingenuity. As we venture forward, the horizons of 3D printing are boundless, fueled by the collaboration of minds from diverse disciplines. The potential applications are as vast as the fields it embraces, promising transformative solutions to global challenges. Through the interplay of biology, engineering, design, medicine, and more, 3D printing is weaving a tapestry of innovation that has the power to reshape industries, enrich lives, and define the future of human progress.

References

1. Crump, S.S. (1992). Apparatus and method for creating three-dimensional objects. Google Patents.
2. Hiemenz, J. (2011). *3D Printing with FDM: How it Works*, vol. 1, 1–5. Stratasys Inc.
3. Go, J., Schiffres, S.N., Stevens, A.G. et al. (2017). Rate limits of additive manufacturing by fused filament fabrication and guidelines for high-throughput system design. *Additive Manufacturing* 16: 1–11.
4. Go, J. and Hart, A.J. (2017). Fast desktop-scale extrusion additive manufacturing. *Additive Manufacturing* 18: 276–284.
5. Duty, C.E., Kunc, V., Compton, B. et al. (2017). Structure and mechanical behavior of Big Area Additive Manufacturing (BAAM) materials. *Rapid Prototyping Journal* 23 (1): 181–189.
6. Levenhagen, N.P. and Dadmun, M.D. (2017). Bimodal molecular weight samples improve the isotropy of 3D printed polymeric samples. *Polymer* 122: 232–241.
7. Ouassil, S.E., El Magri, A., Vanaei, H.R. et al. (2023). Investigating the effect of printing conditions and annealing on the porosity and tensile behavior of 3D-printed polyetherimide material in Z-direction. *Journal of Applied Polymer Science* 140 (4): e53353.
8. Peng, F., Jiang, H., Woods, A. et al. (2019). 3D printing with core–shell filaments containing high or low density polyethylene shells. *ACS Applied Polymer Materials* 1 (2): 275–285.
9. Hart, K.R., Dunn, R.M., and Wetzel, E.D. (2020). Tough, additively manufactured structures fabricated with dual-thermoplastic filaments. *Advanced Engineering Materials* 22 (4): 1901184.
10. Koker, B., Ruckdashel, R., Abajorga, H. et al. (2022). Enhanced interlayer strength and thermal stability via dual material filament for material extrusion additive manufacturing. *Additive Manufacturing* 55: 102807.

2

Potential of 3D Printing from Lab to Industry

Zohreh Mousavi Nejad[1,2], *Nicholas J. Dunne*[1,2,3,4,5,6,7,8,9,10], *and Tanya J. Levingstone*[1,2,3,4,5]

[1] *School of Mechanical and Manufacturing Engineering, Dublin City University, Dublin, Ireland*
[2] *Centre for Medical Engineering Research, School of Mechanical and Manufacturing Engineering, Dublin City University, Dublin, Ireland*
[3] *Advanced Manufacturing Research Centre (I-Form), School of Mechanical and Manufacturing Engineering, Dublin City University, Dublin, Ireland*
[4] *Biodesign Europe, Dublin City University, Dublin, Ireland*
[5] *Advanced Processing Technology Research Centre, Dublin City University, Dublin, Ireland*
[6] *School of Pharmacy, Queen's University Belfast, Belfast, UK*
[7] *School of Chemical Science, Dublin City University, Dublin, Ireland*
[8] *Advanced Materials and Bioengineering Research Centre (AMBER), Trinity College Dublin, Dublin, Ireland*
[9] *Department of Mechanical and Manufacturing Engineering, School of Engineering, Trinity College Dublin, Dublin, Ireland*
[10] *Trinity Centre for Biomedical Engineering, Trinity Biomedical Sciences Institute, Trinity College Dublin, Dublin, Ireland*

2.1 Introduction

The three-dimensional (3D) printing process, also called additive manufacturing (AM), is a method for creating 3D solid objects from digital files. In the era of Industry 4.0, 3D printing is being adopted by more and more companies to optimize their business processes. But like any other relatively new technological advancements, when it comes to 3D printing, a transition from laboratory to industry, and from academic reports to successful commercialization present challenges. However, this transition to 3D printing technology has already occurred to some extent in some fields.

3D printing technology will expand beyond nanotechnology, bioengineering, and manufacturing different products to incorporate a variety of technologies to improve people's quality of life. The potential of 3D printing spans from research and development to industry, as well as the entire value chain from prototypes to spare parts management. A variety of industries

Industrial Strategies and Solutions for 3D Printing: Applications and Optimization, First Edition. Edited by Hamid Reza Vanaei, Sofiane Khelladi, and Abbas Tcharkhtchi.

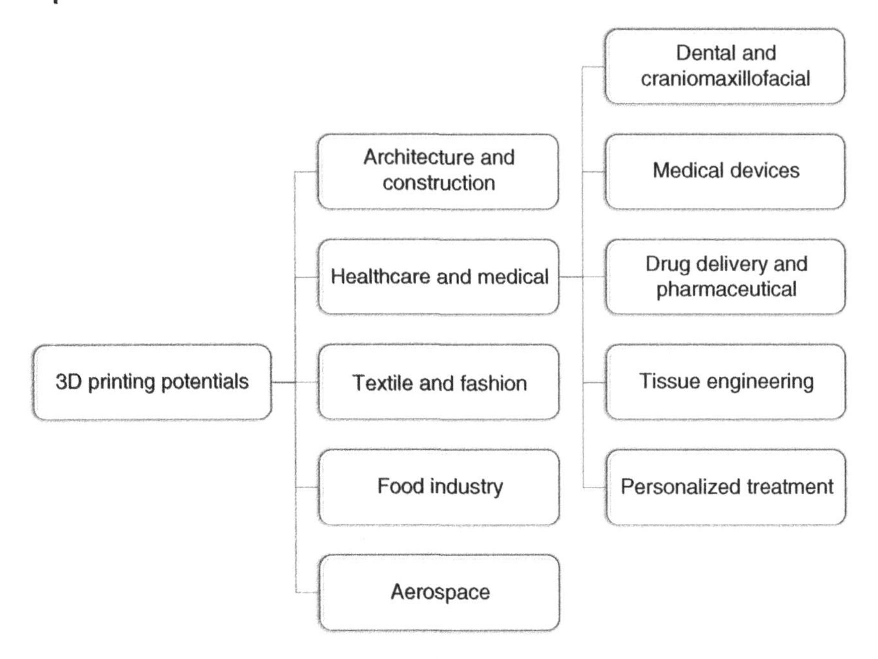

Figure 2.1 The potential of 3D printing in different industries.

are seeking lightweight designs, as reduced material use lowers costs per part and lowers energy consumption. 3D printing enables lightweight designs and reduced material usage in industries. Furthermore, 3D printers can make production faster, which allows companies to become more competitive.

In the field of industrial manufacturing, 3D printing stands out due to its quickness, innovation, and ease of use. Furthermore, 3D printing is typically used to produce prototypes since making changes to them is usually easier and less expensive than resetting tools in a factory. This technology is currently being applied in manufacturing a vast range of objects from the dental prosthesis to aerospace components. The most common potentials of 3D printing technology in different industries are presented in Figure 2.1. Furthermore, the potential of 3D printing and how it can be applied in various industries such as construction, healthcare, fashion, food, and aerospace are discussed in the following sections.

2.2 Architecture and Construction Industry

In recent years, AM technologies have advanced to new levels of sophistication for the building industry which has led to the development and design of new 3D-printable materials, inspired innovative architectural details, and resulted in the development of material-specific printing systems. AM has been renowned for its unique features including its ability to provide the freedom and flexibility of free-form construction, speed, reduced construction costs and waste, and improved safety through innovations in automated construction. In contrast to traditional concrete casting techniques, AM, through the extrusion of concrete, provides greater flexibility and sustainability in terms of creating concrete constructions [1].

Printability and appropriate rheology of concrete are extremely important. It is extremely important to have concrete that is printable and suitable in terms of rheology. In order to prevent damaging printed construction, the concrete should be printed continuously without disconnection, set shortly after extrusion, and maintain a high shape fidelity after setting [2]. Maintaining the shape of concrete struts is more difficult than for other 3D-printed materials such as biomaterials, due to the much heavier weight of concrete compared to bioinks.

Three main approaches have been designed to date for architecture-scale 3D printing systems: (i) the gantry system, (ii) cable-driven systems, and (iii) robotic arm systems. In gantry systems, the print nozzle is moved along the X, Y, and Z axes within a frame structure. The main drawback of this method is that it must be larger than the built structure, so it cannot be used in remote or harsh environments. In a cable-driven system, the printhead hangs on a cable between several fixed points. This eliminates the rigidity issues in gantry systems and is easier to transport. Robotic-arm systems utilize a robot with six axes of freedom and a pump to extrude concrete layer by layer to fabricate a final structure without having to use formwork. Compared with the abovementioned two systems, robotic arm systems are more agile and practical and can create more complex geometries [1].

The application of 3D printing in the construction industry includes bridges, shelters for military applications, social housing, reconstruction of houses damaged by natural disasters, and renovation of buildings [3]. In 2014, DUS Architects started the Canal House project which was Europe's first fully 3D-printed housing project. This project utilized sustainable, biologically derived materials to build an 8-square-meter house using a home-developed 3D printer called "KamerMaker," which could print objects with dimensions of up to $2 \times 2 \times 3.5$ m [3].

In 2016, a 3D-printed office was designed for the United Arab Emirates National Committee as the Dubai Futures Foundation's headquarters by a pioneer architectural 3D printing company called Winsun (Figure 2.2). This 3D-printed office, "Office of the Future," is primarily used as a meeting space for international parties. Air conditioning, telecommunications, water, and

(a) (b)

Figure 2.2 3D-printed office in Dubai called Office of the Future (a) Side view, (b) Front view. Source: Sakin et al. [4] with permission from Elsevier.

electricity are all included in this fully functional 3D-printed office. The parts of this building were printed in China and then transported to Dubai. By using 3D printing technology, this building was constructed with 30–60% less waste and a 50–80% reduction in labor costs [4].

The use of 3D printing technology in the construction of buildings has the potential to increase environmental sustainability. Moreover, 3D printing will enhance creativity of architects, allowing them to overcome existing obstacles and design complex constructions. The use of 3D printing in construction industry could be considered a transformation for the current architecture; however, there is a need to consider the sustainability of both materials and construction methods. 3D printing in construction offers several advantages: lower labor costs as well as materials, environmentally friendly processes with minimal construction and industrial waste, minimal risk of injuries onsite since 3D printers will perform hazardous tasks and at high speed. In addition, there are still many concerns to be addressed, including the possibility of thousands of qualified workers being unemployed as a result of the development of 3D printing technology in construction.

2.3 Healthcare and Medical Industry

3D printing is a promising technology to produce complex structures related to the biomedical and healthcare fields. The application of 3D printing to the medical field once seemed to be a dream, but with the enormous efforts of researchers all over the world, time, and investment, this dream has become a reality. In the healthcare and medical sector, 3D printing is offering more and more applications each year, saving and improving lives in ways never considered before. The technique can be utilized to 3D print computer designs based on anatomical data from patients. This can enable the custom design and manufacture of patient-specific implants and medical devices. Because of the highly accurate nature of 3D printing, this method is ideal for making customized items such as surgical guides, dental molds, and implants for patients, as well as scaffolds for tissue engineering (TE), diagnostic platforms, and drug delivery systems.

A variety of 3D printing technologies, such as extrusion-based 3D printing, selective laser sintering (SLS), and inkjet printing, have been successfully employed and have shown great potential in personalized medicine [5]. Some of the advantages of this technology, such as the ability to build complex geometries that perfectly fit the anatomy of the patient, allow for a wide range of medical applications. These applications range from surgical planning tools to polypills that combine several medications within one single tablet or capsule. The following are the main direct applications of 3D printing in this field.

2.3.1 Dental and Craniomaxillofacial

3D printing is one of the most commonly used technologies in dentistry to create patient-specific dental molds, surgical guides, scaffolds, and prostheses for the repair of complex defects resulting from trauma, cancer, and congenital defects that require a precise anatomical structure [6]. 3D printing technology is well adapted to dentistry since 3D modeling and imaging technologies, such as cone beam computed tomography (CBCT), are widely used in this field. Combining these imaging techniques with computer-aided design/computer-aided manufacturing (CAD/CAM) processes enables the 3D printing of complex implants and devices for dentistry applications. It is anticipated that rapid development within this field could be achieved over the coming years.

Various applications can be obtained by using 3D printing in dentistry, including the fabrication of drill guides for dental implants, the production of craniomaxillofacial and dental implants, the fabrication of provisional dental prosthesis and orthosis, and the manufacture of the covering and frameworks for dental restorations. Furthermore, CBCT, which is an important technology in dentistry, has become more readily available in recent years. CBCT has revolutionized implant dentistry and endodontics through improved diagnosis and treatment. Unlike standard CT imaging, which provides primarily 2D cross-sectional images, CBCT offers the advantage of producing high-quality volumetric "image" data. This data can be sent directly to a 3D printer before surgery, enabling the creation of detailed replicas of the patient's jaws. This approach facilitates a precise review of any anatomy, especially complicated anatomy, and allows for thorough surgical approaches to be practiced or meticulously planned before the actual surgery, enhancing the overall precision and success of dental procedures. As used in a surgical environment, dental implant drill guides and similar engineering tools must be precise and durable as, well as serializable or disinfectable. Drill guides are becoming more common in implant dentistry because they allow the transfer of a virtual 3D plan to the operative site, which can be considered as a bridging tool between virtual plans and actual patients [7].

Additionally, 3D printing technology can be used in dentistry to create dental implants with porous or rough surfaces, which enhance the interaction between the implant and surrounding tissue. However, traditional fabrication methods require extra post-processing steps and costs to make the implant rough or porous.

2.3.2 Medical Devices

In recent years, 3D printing technology has found its place in the field of medical equipment manufacturing. An important potential application for 3D printing in the medical device industry is the production of surgical guides. Along with the use of 3D-printed surgical guides, which have been established for over 10 years, they have also been applied in spinal surgery, neurosurgery, maxillofacial surgery, and orthopedic surgery to provide accurate implant placement and orientation during drilling.

As a result of a clinical study conducted by Zhang et al., an accuracy-improved method was developed for acetabular cup prosthesis implantation in hip replacements. After a 1-year follow-up for patients in whom the navigation template was employed as a surgical guide, there were notably reduced disparities ($1.6° \pm 0.4°$, $1.9° \pm 1.1°$) from the specified abduction angle of 45° and anteversion angle of 18° when compared to individuals in the control group ($5.8° \pm 2.9°$, $3.9° \pm 2.5°$), and this difference was statistically significant ($p < 0.05$) [8]. In another clinical study, critical parameters like operation time, blood loss, and alignment of 40 patients who underwent patient-specific total knee arthroplasty (TKA) were compared to values from a matched control group of patients who were operated on using the conventional intramedullary alignment technique. This study demonstrates that the developed patient-specific method improved alignment accuracy and resulted in a 10-minute reduction in operation time and 60 mL decrease in blood loss compared to the conventional intramedullary alignment technique [9].

Another patient-specific positioning guide was designed by Putzier et al. using 3D printing technique that permits an accurate and safe implantation of pedicle screws in patients with severe scoliosis. A CT-based 3D model of the thoracic and lumbar spine was used to create navigational templates and models for all vertebrae to be instrumented in four patients with severe scoliosis. By designing guides based on patient anatomy, the device was optimized for coupling with the spine, enhanced stability, and improved user-friendliness during screw placement. In addition, intraoperative challenges and opportunities for device and process improvements were documented. A postoperative assessment of screw positions and a check for cortical violation were performed using CT scans. A total of 76 pedicle screws were implanted, with 56 placed in the thoracic region and 20 in the lumbar region. Notably, 84% of these pedicle screws were complete intrapedicular placement, while an impressive 96.1% maintained a distance of less than 2 mm from the cortical boundary, as ascertained through CT scans. There were no indications of incorrect screw placement or pedicle violation near neurovascular structures in the scans. Additionally, no complaints were reported about screw-related complications after surgery [10].

2.3.3 Drug Delivery and Pharmaceutical

3D printing using drug printing is potentially an innovative method for producing patient-specific medicine depending on the composition and dosage required for the patient. Apart from its low cost, the technology can also be used with most pharmaceutically accepted polymers, such as polylactic acid (PLA), cellulose, and methacrylate, to create personalized oral dosage forms [5]. Printing drugs offers the possibility of creating safer, more effective drugs that are tailored to the individual. Additionally, drug printing offers the option of choosing color, shape, and design, which can reduce a child's resistance to taking medication and make medicine more appealing to them [11].

There is an increase in the number of people taking multiple pill-based medications for several conditions during the day, especially as the population ages and chronic diseases become more prevalent. It can be a hassle to take the right pill at the right time, and it can even be dangerous if you make a mistake. Taking a single pill a day that delivers all the essential medications at the right time and in the appropriate dose would make life easier and safer for patients. This may be possible with a polypill, which is a pill form that combines multiple active pharmaceutical ingredients. A commercially available polypill (Polycap™), containing ramipril, aspirin, atenolol, simvastatin, and hydrochlorothiazide, has been shown to reduce multiple cardiovascular risk factors. The process of making personalized polypills with conventional powder compaction is very expensive since they are produced in small quantities – often for a single patient. 3D printing could provide a much cheaper way to make polypills, in which thin layers of materials are built up to build a final product.

Ghanizadeh–Tabriz et al. used fusion deposition modeling (FDM), an extrusion-based 3D printing technology, to design and fabricate a bilayer tablet made up of rifampicin and isoniazid for treating tuberculosis. Rifampicin was formulated in hypromellose acetate succinate matrix to release drugs in the upper intestine (alkaline conditions), while isoniazid was formulated in hydroxypropyl cellulose matrix to release drugs in the stomach (acidic conditions). By minimizing rifampicin degradation under acidic conditions, this design may offer a better clinical efficacy and minimize drug–drug interactions [5].

In another study conducted by Robles–Martinez et al., a polypill combining six drugs, including paracetamol, caffeine, naproxen, chloramphenicol, prednisolone, and aspirin, was printed using stereolithography (SLA) 3D printing. Raman microscopy was used to study drug distribution, which showed that even though separate layers were printed successfully, some drugs diffused throughout the layers, based on whether they were amorphous or crystalline. The printing constructs showed excellent physical properties, and different inclusion materials allowed for distinct drug release profiles for the six actives in dissolution tests. In this study, SLA printing is demonstrated for the first time as a versatile platform for multidrug therapy production, facilitating a new era of personalized polypills [12].

This new form of drug delivery has already undergone a great deal of research, but it is still a long way from being approved by regulators. After passing clinical trials, these approvals can take up to five years. In the future, patients with multiple conditions might just take one 3D-printed pill a day instead of complex schedules of multiple pills. Perhaps in a few years, people will be able to get prescriptions for custom-made polypills manufactured on a 3D printer at their local pharmacies.

2.3.4 Tissue Engineering

The goal of TE is to stimulate tissue regeneration by using porous, biocompatible, biodegradable scaffolds embedded with cells and growth factors and implanting them back into the body to rejuvenate the tissue. Scaffolds have an important role in providing a suitable matrix for cell growth and proliferation, cell signaling, sufficient space for ECM secretion and remodeling, and contributing to injured tissue healing. Thus, the design of a 3D scaffold at macro, micro, and nano levels is important in terms of providing the required structural features, cell-scaffold interaction, and facilitating nutrient transport. A scaffold's macrostructure is the overall shape of the scaffold, which can vary depending on the patient and anatomical features. The microstructure of a scaffold is determined by its porosity, pore interconnectivity, and pore size. The scaffold nanostructure determines the surface features that play a role in cell attachment, proliferation, and differentiation. It would be possible to control the micro and macro structure of the scaffold by 3D printing, and you could easily modify the nanostructure of the scaffold by post-processing [13]. One example of 3D printing potential application in TE is 3D printing synthetic skin for transplantation to burn patients. 3D-printed skin is also applicable to test chemical, pharmaceutical, and cosmetic products. Attempts to imitate human heart valves, bones, teeth, and cartilage are other examples of the use of 3D printing in TE.

The application of 3D printing in the development of new materials for the regeneration of hard tissues, especially bone, is characterized by a wealth of research, reflecting its dynamic and rapidly evolving nature. Wang et al. fabricated a composite scaffold containing PLA and nanohydroxyapatite (nHA) using FDM technology for bone regeneration application. In addition to the physicochemical properties, the in vitro and in vivo osteogenic ability of the PLA/nHA scaffold were evaluated on bone marrow mesenchymal stem cells and a rabbit model, respectively. The results demonstrated that the PLA/n-HA composite was highly printable, and the mechanical strength of the printed composite scaffold was desirable. In addition, the 3D-printed PLA/nHA scaffold had superior biocompatibility and osteogenic induction properties compared to pure PLA scaffolds, making it a promising candidate for large-scale bone regeneration [14].

3D bioprinting is an exciting new tool with the potential to allow simultaneous positioning of biomaterials and cells in a layer-by-layer manner to build a 3D structure that can morphologically and structurally reproduce complex biological tissue. An interesting study by Maiullari et al. designed a 3D multicellular structure composed of human umbilical vein endothelial cells (HUVECs) and induced pluripotent cell-derived cardiomyocytes (iPSC-CMs) for a cardiac tissue model (Figure 2.3). Hydrogel strands containing alginate and PEG-Fibrinogen were used to encapsulate cells, which were then extruded through a microfluidic printhead enabling precise 3D spatial deposition while maintaining high printing fidelity and resolution. In vivo, implantation showed that the 3D-printed blood vessel-like structures integrated better with the host's vessels [15].

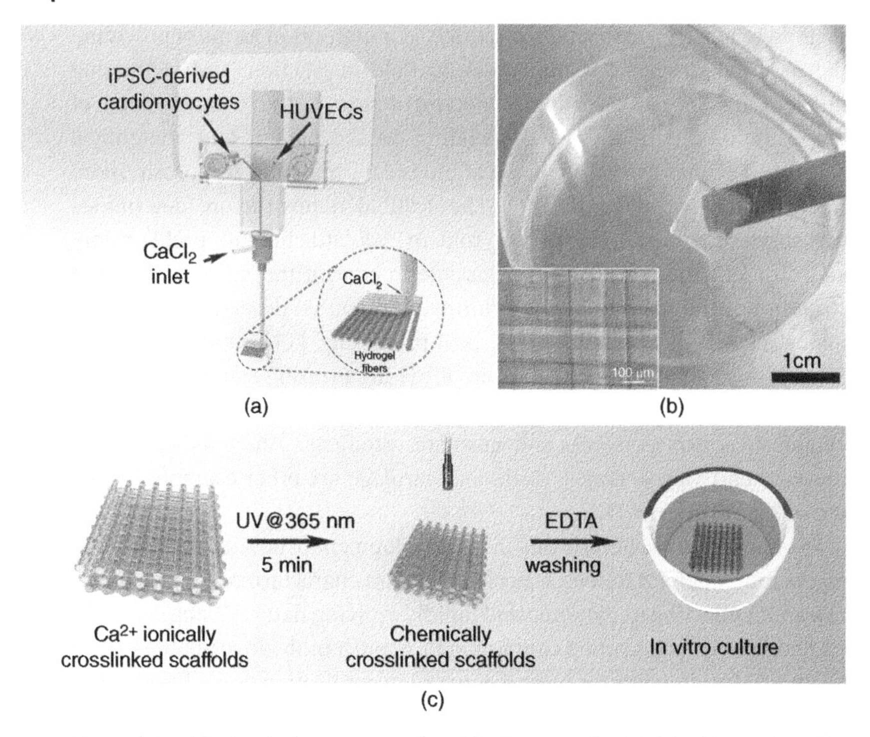

Figure 2.3 3D bioprinting process of multicellular scaffold. (a) An illustration of the microfluidic printing head used to bioprint cardiomyocytes and HUVECs simultaneously. (b) 10-layer thick bioprinted cellularized construct. An image of the same construct at a higher magnification can be found in the box below. (c) This diagram illustrates the process involved in fabricating PEG-Fibrinogen-based scaffold before undergoing in vitro culture. Source: Reproduced with permission from Maiullari et al. [15]/Springer Nature/CC BY 4.0.

In addition to its direct role in TE, 3D printing can also play an indirect role. In an indirect way, EzEldeen et al. utilized 3D printing to fabricate root-canal shape scaffolds with chitosan and gelatine. This study used 3D printing technology to fabricate root-canal molds, followed by freeze-drying to produce 3D scaffolds [16]. Furthermore, within the domain of 3D printing, machine learning (ML) has emerged as a powerful tool in recent years, particularly when applied to optimize various aspects of the process. This is especially relevant in the context of TE applications, where precise calibration of 3D printers is essential for improving the accuracy and consistency of printed structures. This precision plays a pivotal role in creating tissue-engineered constructs with intricate designs, underscoring the crucial interplay between ML and 3D printing in advancing the field of TE [17].

2.3.5 Personalized Treatment

3D printing is a technology that provides pharmaceutical and medical companies with a great opportunity not only to develop more specific drugs but also to quickly manufacture medical implants and enhance the way doctors and surgeons plan surgery. 3D printing can also be used for planning preoperative procedures and personalizing treatments. This will result in a multistep procedure that integrates clinical and imaging data to enable the most effective treatment option to be determined. Patient-specific preoperative planning has been shown to lead to shorter operating room times, reduced complications, shorter postoperative stays, and lower medical costs. Cross-sectional imaging combined with 3D printing technology led to an accurate presurgical plan by providing the surgeon with a 3D model of the patient's anatomy. As a result, a deeper understanding of the complex anatomy of each case can be achieved. Furthermore, 3D printing gives a high degree of precision in selecting the size of prosthetic components before they are implanted [11].

Custom-designed implants or prostheses are one of the most promising 3D printing applications in medical devices because of the advantage of being able to create implants for complex clinical cases. In knee or hip surgery, general implants can be produced to meet the needs of a large number of patients. However, for some applications that treat complex anatomical structures in patients, the use of common implants is not recommended, and custom implants are required for satisfactory results.

In recent years, implants have been widely used for tissue reconstruction after pelvic tumor resection. However, currently, these implants are associated with tumor recurrence, poor implant fit, defects after tumor resection, and loosening or failure of the implant. The use of 3D-printed osteotomy guides for postoperative reconstruction after pelvic tumor resection results in more accurate preoperative design, complete tumor resection, and better anatomical conformance and prosthesis stability. Applying 3D printing to create a surgical guide and personalized prosthesis for pelvic tumor resection improves prosthesis–patient matching reduces surgical trauma, reduces surgical time, and reduces patient postoperative time.

2.4 Textile and Fashion Industry

3D printing is becoming more popular in the fashion industry as some fashion designers try to create wearable garments made with AM. For the past decade, researchers, textile engineers, and fashion designers have been working to apply 3D printing technology in their respective fields. 3D printing is extremely beneficial in the production of fiber-reinforced composites and has proven to significantly reduce manufacturing time and costs. However, the application of this technique in the production of everyday clothing is still a niche.

Textile apparel and fashion accessories require flexibility, strength, expandability, and resilience to ensure customer comfort and adapt to body movements. Compared to traditional clothing techniques, 3D-printed clothing offers consumers a greater degree of customization. Thus, 3D printing technology offers the possibility of customizing clothing according to the shape and size of an individual. It may be even possible to scan the human body and adjust personalized clothing accordingly. Wearable smart textiles and fashion products with small, lightweight, and sensitive trackers have attracted attention, and researchers are attempting to design and manufacture electronic textiles using 3D printing technology.

Textile and apparel sustainability trends underscore the need to ensure the sustainability of 3D printing technology by reducing waste generation and reducing transportation and energy costs. Researchers have also tried to design and manufacture costume jewelry using durable materials and producing zero waste [18]. During the Paris Fashion Week Spring 2013, a 3D-printed dress was demonstrated for the show "Voltage" by fashion designer Iris van Herpen in collaboration with Professor Neri Oxman [19]. This set of 3D-printed skirt and cape was produced with Stratasys' multi-material Objet Connex technology, which allows a variety of materials to be printed simultaneously. Using a variety of hard and soft materials in both pieces gave the garments movement and texture [19].

Wearable textiles with trackers and sensors allow the self-monitoring of physical activity and health through visual and numerical displays of the collected data. Furthermore, electronic clothing helps athletes monitor their performance and health by recording body temperature, sweat, and muscle movement during all types of sporting activities. A 3D-printed add-on part may serve a special function based on the end-use application in addition to providing comfort and aesthetics. There is a variety of end-use applications for these add-on parts, including athletic shoes, e-textiles, sports gear, medical and industrial applications, and textile reinforcement.

In the past, 3D printing has been restricted to materials such as polyvinyl alcohol (PVA), PLA, and acrylonitrile butadiene styrene (ABS). These synthetic materials and the available 3D printing techniques did not meet the intricate demands of 3D-printed fashion items, and the fabrics produced were stiff and rigid and would likely be too uncomfortable for use in clothing. Today, designers and manufacturers have greater flexibility thanks to the availability of diverse materials and filaments. Textile and fashion designers now commonly utilize materials such as cellulose composites (cellulose acetate/acetic acid), polyethylene terephthalate (PET), polyamide, PVA, ABS, PLA, and ceramics. These materials now find application not only in traditional fabrics but also in garment components and functional textiles, such as cutting-edge electronic textiles (e-textiles) that combine electronics and textiles. This expanded material palette not only enhances creative possibilities but also opens up new opportunities for 3D-printed fashion innovation.

2.5 Food Industry

The application of 3D printing technology to the food sector can bring a number of advantages, such as the customization of food designs, the personalization and digitization of nutrition, the simplification of supply chain, and the expansion of the available food sources. Food designers, chefs, and nutritionists can utilize this technology to create complex and fantastic food designs that are not achievable with traditional molds or manual labor. It can also be used to decorate solid edible substrates with confectionery shapes and colorful images. Furthermore, 3D food printing allows us to digitize and personalize individuals' nutritional and energy requirements based on their physical and nutritional status.

There are currently four 3D printing techniques available in the food sector, including extrusion-based printing, binder jetting, SLS, and inkjet printing. Extrusion-based printing is more common in food printing, as it is usually used to extrude hot chocolate or soft materials like mashed potatoes, dough, or meat puree. Recently, researchers have made great efforts to apply 3D printing of food in the food industry. Although there are still many challenges that prevent this technology from being widely used in the food sector due to numerous reasons, including process productivity, printing precision and accuracy, and production of colorful, multiflavor, multitexture objects. A benefit of 3D printing is the ability to create exquisite and attractive structures in edible products, which increases the interest and appetite of consumers. In food printing, the accuracy and precision of printing will depend on several factors: material properties (such as rheological properties and particle size), process parameters (such as print speed and distance), and post-processing methods (such as baking, cooking, and frying) [20].

In a clinical study conducted by Zhang et al., with the aid of food 3D printing, a perceptual illusion was created to demonstrate how to control perceived satiety given a defined calorie amount. They present a 3D printing system called FoodFab, which allows users to change a food's internal structure via two critical 3D printing parameters: infill density and infill pattern. Using 30 participants in this experiment, they examined how these parameters affected chewing time, which affects people's feelings of satiety [21].

Another study was undertaken to determine whether cereal-based food structures containing probiotics could be manufactured using 3D printing. During the 3D printing process, the printability of dough formulations with different water contents, wheat flour types, and calcium caseinate levels was evaluated. Different printing patterns of dough are shown in Figure 2.4. The study showed that the dynamic rheological properties and microstructure of dough were influenced by its composition. In terms of both printing results and post-printing stability, dough formulations with higher viscoelastic modulus, yield stress, and complex viscosity achieved excellent printing results [22].

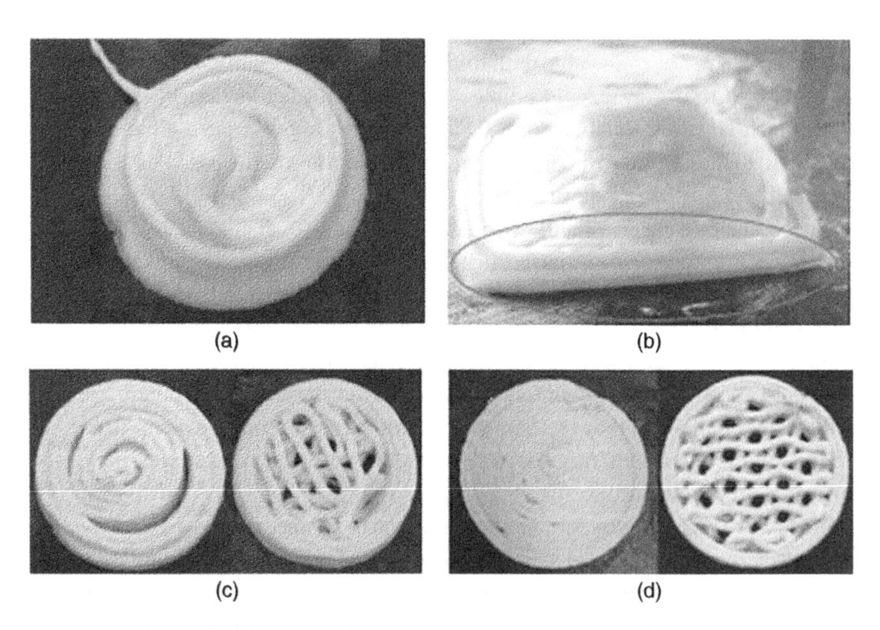

Figure 2.4 3D-printed doughs with different formulations: (a) flour with protein ratio of 7.2% w/w and no calcium caseinate; (b) flour with protein ratio of 12% w/w and no calcium caseinate; (c) flour with protein ratio of 7.2% w/w, 1.17 g calcium caseinate, and nozzle size 1.6 mm; (d) flour with protein ratio of 7.2% w/w, 1.17 g calcium caseinate, and nozzle size 1.2 mm. Flour and water content were 39 and 30 g, respectively, for all the doughs. Source: Reproduced with permission from Zhang et al. [22]/Elsevier.

2.6 Aerospace Industry

3D printing is used by the aerospace industry for prototyping, testing, and producing parts for use in the field. Laser-based 3D printing technologies such as selective laser melting (SLM), direct metal laser sintering (DMLS), and laser metal deposition (LMD) are the most widely used for manufacturing aerospace components [23]. Aerospace applications prove to be the most challenging application of 3D printing technology. Rockets and satellites are exposed to extreme temperatures, and they are subjected to harsh structural, acoustic, and vibration loads [24]. Furthermore, aerospace applications need to address long-term aging and fatigue issues. Aircraft manufacturers should also consider repairing and reworking space-related applications. These requirements increase the need for high-performance alloy materials that can meet a wide range of simultaneous requirements. The aerospace industry has two major requirements: lightweight components and high safety. 3D printing can help reduce weight by producing complex, mesh-like structures with a small number of connections and complex shapes [23].

The aerospace industry demands components that are not only lighter but also stronger and more durable. Currently, 3D printing technology is creating new ways to meet these challenges. The application areas include rapid prototyping using plastic and metal in the design phase, making of dies, molds, and tools for mass production, and direct manufacturing of complex-shaped metal parts. Repairing damaged parts instead of scrapping them or having them replaced is also an option. LMD technology has been shown to be the most effective technique for the repair of aerospace components. This method involves applying metal powder directly to the damaged part of the part and applying laser curing to restore the part to its original strength.

Currently, aerospace parts are produced by machining, forging, and other conventional processes. In conventional manufacturing, an excessive amount of highly expensive material is wasted, and only 5% of it is used in the finished part. The material is utilized to the fullest extent possible in 3D printing. A variety of companies have used 3D printing technology in the aerospace sector. There are a number of examples, such as Martin F-35 aircraft and wing leading edges manufactured by GKN Aerospace, dozens of brackets made out of titanium alloy by Lockheed Martin, and a 3D-printed Inconel pressure vent used by NASA in the flight test of Orion.

In aerospace engines, the operating temperature directly affects fuel efficiency. With 3D printing, high-temperature materials such as nickel alloys and intermetallic materials that are difficult to machine and cast can be processed. With the flexibility of the process, designers can design sophisticated component assemblies with a variety of shapes, compositions, structures, and properties [23].

2.7 Conclusions and Future Perspectives

Three decades ago, 3D printing was introduced as a versatile manufacturing technology. A wide range of items, from houses to space equipment, can be quickly and efficiently manufactured using the technique. The use of 3D printing has expanded rapidly from other sectors into medicine in the last ten years, initially in orthopedic surgery, dentistry, and maxillofacial surgery. As a revolutionary advanced manufacturing technology, 3D printing has demonstrated its potential to improve architectural design, facilitate more affordable, accessible, and personalized medical treatment, produce e-textiles and 3D-printed clothes, as well as to manufacture lightweight, high-temperature resistant, and durable aerospace components. It is currently used in a wide variety of fields due to its cutting-edge nature. It is therefore clear that 3D printing will continue to grow and innovate for the benefit of a wide range of industries. However, it is important to acknowledge that there are still some limitations to 3D printing technology that need to be addressed. One significant challenge is the need for the development of new materials that can meet the diverse and often strict requirements of different industries. With continued advancement of available 3D printable materials, continued improvement in 3D print resolution, and reduction in equipment costs, it is anticipated that the use of 3D printing will continue to expand rapidly in the coming years. These developments will bring with them a wide range of innovations across an ever-diversifying range of industries.

References

1 Nazarian, S., Duarte, J.P., Bilén, S.G. et al. (2021). Additive manufacturing of architectural structures: an interplay between materials, systems, and design. In: *Advances in Science, Technology and Innovation*, 111–119. Springer Nature https://doi.org/10.1007/978-3-030-35533-3_15.
2 Biernacki, J.J., Bullard, J.W., Sant, G. et al. (2017). Cements in the 21st century: challenges, perspectives, and opportunities. *Journal of the American Ceramic Society* 100 (7): 2746–2773. https://doi.org/10.1111/jace.14948.

3 Freire, T., Brun, F., Mateus, A., and Gaspar, F. (2021). 3D Printing technology in the construction industry. In: *Advances in Science, Technology & Innovation* (ed. H. Rodrigues, F. Gaspar, P. Fernandes, and A. Mateus), 157–167. Cham: Springer International Publishing https://doi.org/10.1007/978-3-030-35533-3_19.
4 Sakin, M. and Kiroglu, Y.C. (2017). 3D Printing of buildings: construction of the sustainable houses of the future by BIM. *Energy Procedia* 134: 702–711. https://doi.org/10.1016/j.egypro.2017.09.562.
5 Tabriz, A.G., Nandi, U., Hurt, A.P. et al. (2021). 3D printed bilayer tablet with dual controlled drug release for tuberculosis treatment. *International Journal of Pharmaceutics* 593: 120147. https://doi.org/10.1016/j.ijpharm.2020.120147.
6 EzEldeen, M., Moroni, L., Nejad, Z.M. et al. (2023). Biofabrication of engineered dento-alveolar tissue. *Biomaterials Advances* 148: 213371. https://doi.org/10.1016/j.bioadv.2023.213371.
7 Dawood, A., Marti Marti, B., Sauret-Jackson, V., and Darwood, A. (2015). 3D printing in dentistry. *British Dental Journal* 219 (11): 521–529. https://doi.org/10.1038/sj.bdj.2015.914.
8 Zhang, Y.Z., Chen, B., Lu, S. et al. (2011). Preliminary application of computer-assisted patient-specific acetabular navigational template for total hip arthroplasty in adult single development dysplasia of the hip. *The International Journal of Medical Robotics and Computer Assisted Surgery* 7 (4): 469–474. https://doi.org/10.1002/rcs.423.
9 Boonen, B., Schotanus, M.G.M., and Kort, N.P. (2012). Preliminary experience with the patient-specific templating total knee arthroplasty. *Acta Orthopaedica* 83 (4): 387–393. https://doi.org/10.3109/17453674.2012.711700.
10 Putzier, M., Strube, P., Cecchinato, R. et al. (2017). A new navigational tool for pedicle screw placement in patients with severe scoliosis: a pilot study to prove feasibility, accuracy, and identify operative challenges. *Clinical Spine Surgery* 30 (4): E430–E439. https://doi.org/10.1097/BSD.0000000000000220.
11 Aimar, A., Palermo, A., and Innocenti, B. (2019). The role of 3D printing in medical applications: a state of the art. *Journal of Healthcare Engineering* 2019: 5340616. https://doi.org/10.1155/2019/5340616.
12 Robles-Martinez, P. et al. (2019). 3D Printing of a multi-layered polypill containing six drugs using a novel stereolithographic method. *Pharmaceutics* 11 (6): 274. https://doi.org/10.3390/pharmaceutics11060274.
13 Chia, H.N. and Wu, B.M. (2015). Recent advances in 3D printing of biomaterials. *Journal of Biological Engineering* 9 (1): 4. https://doi.org/10.1186/s13036-015-0001-4.

14 Wang, W., Zhang, B., Li, M. et al. (2021). 3D printing of PLA/n-HA composite scaffolds with customized mechanical properties and biological functions for bone tissue engineering. *Composites Part B: Engineering* 224: 109192. https://doi.org/10.1016/j.compositesb.2021.109192.

15 Maiullari, F., Costantini, M., Milan, M. et al. (2018). A multi-cellular 3D bioprinting approach for vascularized heart tissue engineering based on HUVECs and iPSC-derived cardiomyocytes. *Scientific Reports* 8 (1): https://doi.org/10.1038/s41598-018-31848-x.

16 Ezeldeen, M., Loos, J., Mousavi Nejad, Z. et al. (2021). 3D-printing-assisted fabrication of chitosan scaffolds from different sources and cross-linkers for dental tissue engineering. *European Cells & Materials* 40: https://doi.org/10.22203/eCM.v041a31.

17 Al-Kharusi, G., Dunne, N.J., Little, S., and Levingstone, T.J. (2022). The role of machine learning and design of experiments in the advancement of biomaterial and tissue engineering research. *Bioengineering* 9 (10): https://doi.org/10.3390/bioengineering9100561.

18 Chakraborty, S. and Biswas, M.C. (2020). 3D printing technology of polymer-fiber composites in textile and fashion industry: a potential roadmap of concept to consumer. *Composite Structures* 248: 112562. https://doi.org/10.1016/j.compstruct.2020.112562.

19 Oxman, N., Anthozoa: cape & skirt, 3-D printed dress in collaboration with Iris Van Herpen | by Neri Oxman. http://neri.media.mit.edu/projects/details/anthozoa (accessed Sep. 05, 2023).

20 Liu, Z., Zhang, M., Bhandari, B., and Wang, Y. (2017). 3D printing: printing precision and application in food sector. *Trends in Food Science & Technology* 69: 83–94. https://doi.org/10.1016/j.tifs.2017.08.018.

21 Lin, Y.J., Punpongsanon, P., Wen, X. et al. (2020). FoodFab: creating food perception illusions using food 3D printing. In: *Proceedings of the 2020 CHI Conference on Human Factors in Computing Systems*, 1–13. Honolulu HI USA: ACM https://doi.org/10.1145/3313831.3376421.

22 Zhang, L., Lou, Y., and Schutyser, M.A.I. (2018). 3D printing of cereal-based food structures containing probiotics. *Food Structure* 18: 14–22. https://doi.org/10.1016/j.foostr.2018.10.002.

23 Joshi, S.C. and Sheikh, A.A. (2015). 3D printing in aerospace and its long-term sustainability. *Virtual and Physical Prototyping* 10 (4): 175–185. https://doi.org/10.1080/17452759.2015.1111519.

24 Schiller, G.J. (2015). Additive manufacturing for aerospace. In: *IEEE Aerospace Conference*, 1–8. https://doi.org/10.1109/AERO.2015.7118958.

3

Applicable Materials and Techniques in 3D Printing

Saeedeh Vanaei and Mohammad Elahinia

Department of Mechanical, Industrial and Manufacturing Engineering, University of Toledo, Toledo, OH, 43606, USA

3.1 Introduction

3D printing, often referred to as additive manufacturing, has emerged as a revolutionary technology with the potential to transform various industries. This method involves creating three-dimensional objects layer-by-layer, directly from digital designs, offering unprecedented freedom in terms of geometry and design complexity. The allure of producing intricate parts with reduced costs has captured the attention of both researchers and manufacturers, spurring rapid advancements in the field. The evolution of 3D printing has been marked by significant progress across a range of aspects. One of its key advantages lies in its ability to produce objects with complex geometries that would be extremely challenging or even impossible to achieve through traditional manufacturing methods.

This characteristic has fueled interest across industries where intricately designed components are essential, such as aerospace, automotive, energy, and medical devices. Furthermore, the diverse materials that can now be utilized in 3D printing have broadened its application horizons. This technology has transcended the boundaries of just plastics and polymers, now encompassing metals, ceramics, and composites. This method versatility has opened doors to tailor-made solutions for specific applications within various sectors. Despite the numerous advantages that 3D printing brings to the table, there are still aspects that necessitate further exploration and understanding. The engineering of material properties in the context of 3D printing is a particularly crucial area. The interplay between process parameters, such as temperature, layer height, print speed, and material

Industrial Strategies and Solutions for 3D Printing: Applications and Optimization, First Edition. Edited by Hamid Reza Vanaei, Sofiane Khelladi, and Abbas Tcharkhtchi.

composition, has a direct impact on the final mechanical, thermal, and chemical properties of the printed object.

The concept of rapid solidification and reheating is particularly pertinent in understanding how the material properties evolve during the printing process. As each new layer is added, the freshly deposited material rapidly cools and solidifies. This intricate cooling process influences the microstructure and material properties. Upon subsequent layers being added, reheating occurs as the new material is deposited on top. This combination of rapid solidification and reheating phases demands a deep comprehension of material behavior to ensure the desired properties are achieved consistently. A profound understanding of these aspects is crucial to fully harness the potential of 3D printing. It not only aids in producing components with superior performance characteristics but also minimizes material wastage, which is a notable advantage of additive manufacturing. Fast prototyping, which allows for swift iteration and refinement of designs, is another key benefit that accelerates innovation in various industries.

The evolution of 3D printing has brought about remarkable advancements in manufacturing technology, offering unparalleled design freedom, reduced material waste, and rapid prototyping capabilities. While its advantages are evident, there is a pressing need for further research to unravel the complexities of material engineering in the context of 3D printing. A comprehensive grasp of how process parameters influence material properties, especially during rapid solidification and reheating, will empower industries to fully capitalize on this transformative technology [1]. 3D printing encompasses a diverse array of techniques, each tailored to specific material feedstock forms, such as powder, filament, wire, and sheets. The choice of technique is a crucial decision influenced by the desired application and the specific properties required for the end product.

This chapter delves into the intricate world of materials employed in 3D printing, followed by an exploration of the various 3D printing processes. These processes are analyzed in conjunction with the materials they utilize, and the respective advantages and drawbacks of each technique are scrutinized for comprehensive comparison. The utilization of different material forms is a critical facet of 3D printing. Powder-based techniques involve layering fine particles of material, which are selectively fused together using lasers or other heat sources such as electron beam and plasma arc. Filament-based methods, on the other hand, rely on thermoplastic filaments that are extruded layer-by-layer and solidify upon cooling. Wire- and sheet-based approaches incorporate metal wires or sheets that are incrementally shaped and bonded to create the desired object.

This chapter further embarks on explicating the variety of materials compatible with 3D printing. The gamut spans from traditional polymers and plastics to advanced metals, ceramics, composites, and even biological materials. Each material category offers a distinct set of mechanical, thermal, and chemical properties, rendering them suitable for specific applications across industries.

In summation, the chapter traverses through the intricate interplay of material forms, the diversity of materials themselves, and the spectrum of 3D printing techniques. Integrating these components, not only facilitates a comprehensive understanding of the 3D printing landscape but also empowers decision-makers to choose the most fitting approach for their specific application. This holistic exploration of materials, processes, and their corresponding merits and demerits sets the stage for unlocking the true potential of 3D printing across various industries and applications.

3.2 Materials in 3D Printing

3.2.1 Metals

3.2.1.1 Aluminum Alloys

3D printing of aluminum alloys is utilized in aerospace and automotive applications. The design and manufacturing of porous structures within Al alloys result in lightweight parts with improved mechanical properties. Moreover, these alloys are challenging to machine and shape. As a result, 3D printing of Al alloys is employed to produce parts for the mentioned applications. Another advantage of 3D printing in Al alloys is the control and enhancement of microstructures. Conventionally, microstructure enhancement in Al alloys is achieved by adding chemicals during casting. However, through 3D printing, microstructure refinement is achieved via the rapid cooling process. The challenges associated with LPBF (Laser Powder Bed Fusion) of Al alloys stem from the nature of their powder. The lightweight powder with low laser absorbance at the laser wavelength in LPBF necessitates the use of predominantly cast alloys for 3D printing of aluminum. Among these, AlSi10Mg and AlSi12 are the leading Al alloys [2].

3.2.1.2 Stainless Steel

Stainless steels are iron-based alloys that find extensive use in various applications, including automotive, building structures, and medical devices, owing to their numerous advantages. These advantages encompass weldability, good corrosion resistance, ductility, and relatively low cost. Stainless steel 316L stands as a representative example of such alloys. Unlike aluminum alloys, stainless steel possesses good weldability, rendering it suitable for 3D printing. It is manufactured through powder bed fusion (PBF) and direct energy deposition (DED) [3].

3.2.1.3 Titanium Alloys

Titanium alloys are employed in high-performance applications due to their excellent mechanical properties. Additionally, these alloys stand as leading materials in biomedical applications. However, the manufacturing of these alloys is challenging owing to their low ductility, and the cost of fabricating parts using conventional methods is relatively high. As a result, 3D printing has garnered interest as a means to fabricate titanium alloys, namely Ti6Al4V, Ti6Al7Nb, Ti6.5Al3.5Mo1.5Zr0.3Si, and pure titanium [4, 5].

3.2.1.4 Nickel-based Shape Memory Alloys

Shape memory alloys (SMAs) are a class of smart materials with properties such as super elasticity, corrosion resistance, damping characteristics, and biocompatibility, and have been widely used in applications such as aerospace and medical devices. NiTi and NiTi-based SMAs are the most widely studied and commercially used shape memory alloys due to their functional features namely shape memory and super elasticity. These properties are based on the solid phase transformation between austenite and martensite phase that leads to recoverable deformation of up to 10% upon heating and loading of the material [6, 7]. One of the challenges with this material is its machinability. Therefore, over the past decade, 3D printing of NiTi-based SMAs using powder bed fusion techniques has been extensively studied because of its capability to overcome challenges associated with conventional methods of fabrication such as casting and powder metallurgy [8]. Powder characteristics, 3D printing parameters, and post processing such as heat treatment and surface treatment are important factors in 3D printing of NiTi-based SMAs. Powder properties such as powder size, morphology, and size distribution as well as the composition and Ni/Ti ratio affect the microstructure and properties of the final part. Addition of other elements alters the transformation temperatures, which can be interesting for applications at low or high temperatures. It also results in different precipitations that affects the microstructure and texture of the alloy, and consequently the mechanical properties and thermomechanical response of the material. Process parameters such as laser power, scanning speed, scanning strategy, building orientation, and layer thickness affect the properties and functionality of the printed part [9].

3.2.1.5 Cobalt Chrome Alloys

The primary application of cobalt chrome-based alloys is in dentistry and prosthodontics due to their excellent corrosion resistance, mechanical strength, and biocompatibility. However, the processing of this alloy is challenging owing to its high hardness [10]. Surface characteristics are highly dependent on the processing of the alloy. 3D printing has shown improved surface properties and biocompatibility of cobalt–chromium alloys. Haan et al. [11] achieved enhanced fatigue performance through hot isostatic pressure post-processing of Co-Cr parts fabricated by LPBF.

3.2.2 Polymers

Polymers encompass a range of materials, including thermoplastics, thermosets, elastomers, hydrogels, polymer blends, biological systems, and polymer-based composites. The manufacturing of polymers using 3D printing techniques has been investigated and proven to be beneficial. Polymers are primarily produced through fused deposition modeling (FDM) and selective laser sintering (SLS). The following section will discuss the most commonly used polymers in 3D printing.

3.2.2.1 Polylactide

Polylactide (PLA), known for its high Young's modulus, good printability, and environmentally friendly nature, stands as a favorable material for biomedical applications like tissue engineering scaffolds and drug delivery. PLA takes a prominent role among the primary polymers used in 3D printing, specifically in the Fused Filament Fabrication (FFF) process. This preference is owed to the mentioned properties, as well as its low melting temperature, strong adhesion to the build plate, and reduced delamination susceptibility. Despite these advantages, PLA's 3D printing potential is limited by its low impact strength, solubility in water, and low heat resistance. Consequently, researchers are incorporating other materials to create composites that exhibit enhanced properties [12].

3.2.2.2 Acrylonitrile Butadiene Styrene

Acrylonitrile butadiene styrene (ABS) emerges through the polymerization of acrylonitrile, butadiene, and styrene, with its characteristics shaped by the proportions and interplay of these constituent monomers. ABS offers favorable mechanical attributes, elevated resilience, thermal stability, affordability, and straightforward processing [13].

3.2.2.3 Polyamide

Polyamide, commonly referred to as nylon, stands as a thermoplastic polymer acclaimed for its exceptional characteristics. Yet, utilizing pure polyamide often leads to flawed components marked by warping, distortion, and insufficient stability. These challenges can be addressed by modifying the parameters of 3D printing. Furthermore, the incorporation of polymer blends, where polyamide is combined with amorphous polymers, has demonstrated efficacy in thwarting warping and shrinkage. Introducing composites of polyamide infused with particle reinforcements such as iron (Fe) and aluminum (Al), alongside ceramic particles, has emerged as a viable approach to enhance its performance for welding applications [14].

3.2.2.4 Polycarbonate

Polycarbonate, recognized as a thermoplastic polymer, boasts notable robustness and endurance, commendable resistance to heat, and substantial viscosity. It finds utility across automotive, medical, and optical domains. Polycarbonate exhibits a relatively elevated glass transition temperature (150 °C), retaining its structural integrity at temperatures up to this point [15].

3.2.3 Ceramics

Ceramics, distinguished by attributes like elevated hardness and mechanical strength, are widely employed across electronics, aerospace, biomedical, and tissue engineering sectors. Traditional manufacturing approaches encounter limitations in shaping intricate ceramic structures due to their brittleness and the resultant challenges in machining, including tool wear, cracking, and material defects. The introduction of 3D printing has piqued interest in ceramic 3D printing as a solution to these issues. Both dense and porous ceramic structures, along with ceramic composites, find utility in 3D printing applications. The principal ceramic 3D printing techniques encompass binder jetting and PBF. The selection of a technique is guided by the desired final material properties and the associated pace and expense of each process. In 3D printing methodologies that employ powdered material, such as PBF, ceramic particles serve as the building blocks for fabricating components. In fused deposition filament processes, wherein the material is supplied as filament, ceramic particles are blended with a binder to create the filaments [16].

3.2.4 Composites

Advanced technologies often employ composite materials recognized for their exceptional attributes, including reduced weight. Composite materials are composed of a base matrix, which can be metal, polymer, or ceramic, and this matrix is fortified through the integration of an additional substance. This reinforcing element can manifest as either fibers or powders [17].

3.2.4.1 Fiber Reinforced Composites

Incorporating fibers, typically carbon and glass filaments, can substantially enhance the mechanical characteristics of polymers like PLA, and this practice finds application in 3D printing. Carbon fiber-reinforced polymers are widely adopted in aerospace contexts. Among 3D printing techniques, FDM stands as the prevalent choice for manufacturing fiber-reinforced polymer composites. Common composite configurations encompass carbon/glass fiber within ABS, carbon fiber within PLA, and carbon fiber within nylon. The attributes of the fiber, including its alignment and arrangement, exert an impact on the traits of the printed component, and these factors can be adjusted to achieve compact parts with heightened mechanical attributes [18, 19].

3.2.4.2 Particle Reinforced Composites

The utilization of particle reinforcement within a polymer matrix is a topic of significance due to the cost-effectiveness associated with particles. Various types of material particles can be incorporated based on the desired properties. For instance, the introduction of iron (Fe) and copper (Cu) particles is a common approach to enhance tensile strength, while aluminum (Al) particles are employed to elevate wear resistance. Widely used polymers such as ABS, resin, polycarbonate, polypropylene, and nylon are frequently chosen for these applications. Particle-reinforced polymers hold particular interest as they exhibit potential for structural applications due to the resultant improvements in mechanical characteristics. Furthermore, the incorporation of particles, such as Fe and Cu particles within an ABS polymer matrix, has demonstrated promising outcomes in addressing 3D printing-induced distortions by mitigating thermal expansion.

3.3 Techniques in 3D Printing

There are a variety of classifications for 3D printing methods, based on various criteria such as feedstock, energy source, and build volume. The ASTM 52900:2015 standard categorizes 3D printing techniques into seven groups: binder jetting, directed energy deposition, material extrusion (FDM), powder bed fusion, material jetting, sheet lamination, and vat photopolymerization. These techniques will be discussed in the following sections. At the end of the chapter, a summary of these processes can be found in Table 3.1, detailing the materials used, advantages, limitations, and applications of each technique.

Table 3.1 Summary of different 3D printing techniques, materials, applications, advantages, and drawbacks.

Method	Feedstock	Materials	Advantage	Disadvantage	Applications
FDM (FFF)	Filament	ABS PLA PCL PE	Low cost Moderate melting temperature	Pore formation Delamination	Automotive Biomedical Pharmaceutical Electronic
BJ	Powder	316 L stainless steel Ceramic Glass Inconel Fe alloys	High production rate High resolution Design freedom Low cost	Poor strength	Electronics Gas turbine blades Medical
DED	Powder Wire	Stainless steel Ni, Ti, Co, and Al alloys	High deposition rate Large build volume	Poor resolution Low repeatability Surface roughness	Hip implants Aerospace components Turbomachine components
PBF	Powder	Ti alloys Ni-based superalloys 316L Stainless steel	High resolution Dimension control	Slow deposition rate Small build volume	Aircraft engine Medical devices Gas turbines Actuators

MJ	Polymers	PLA PA ABS	High dimensional accuracy Low surface roughness Multi-material printing	Shrinkage Oxygen pick-up	Dentistry, Biomedical, Electronic, Aviation
ShL	Sheets	Paper Polymers Ceramics	High surface quality finish Low cost	Limitations in highly complex designs Poor dimension accuracy Poor surface finish	Paper industry Electronics
VP	Polymer Polymer-ceramic composites	Biocompatible, durable, flexible, and structural resins Bioinks	Excellent surface Resolution High accuracy High thermal durability	Handling the feedstock	Medical (hearing aids, cell scaffolds, and microneedles) Electronics Jewelry

3.3.1 Fused Deposition Modeling

FDM, also known as FFF, has emerged as a promising technique for fabricating polymers and their composites, with ABS and polylactic acid (PLA) standing as the dominant materials. The feedstock takes the form of a filament, and the produced parts find applications across various sectors, including automotive, biomedical, and electronics. Process parameters within FDM, such as temperature, printing speed, layer thickness, and infill density, have been extensively studied and optimized to achieve enhanced properties [20]. The deposition of molten layers of filaments introduces challenges, notably delamination, which can have adverse effects on mechanical properties [21].

3.3.2 Powder Bed Fusion

PBF stands as a frequently employed technique, particularly for metal fabrication. In this approach, a layer of powder is selectively melted and fused together by an energy source, typically either a laser or an electron beam (the latter is also referred to as electron beam melting). The solidification of the deposited layer follows after the build platform moves downward in accordance with the designated layer thickness. This paves the way for the deposition of the subsequent layer of powder, and this sequence is repeated to ultimately craft the final part through a layer-by-layer process. PBF is versatile, enabling the production of both fully dense and porous structures. However, due to the rapid cooling during PBF, internal flaws may arise within the part. As the deposited layer resolidifies upon the melting of the ensuing powder layer, it may lead to shrinkage or the entrapment of the build chamber shielding gas. Parameters integral to the process – such as laser power, layer thickness, and scanning speed – hold sway over the part's properties. Consequently, adjusting these parameters becomes instrumental in achieving the desired microstructure and properties aligned with the intended application. In comparison to DED, PBF offers a greater capacity for intricate geometries with heightened precision, thereby solidifying its status as the prevalent method in aerospace applications [22].

3.3.3 Direct Energy Deposition

DED stands out as one of the primary 3D printing methods for metal fabrication. The feedstock can take the form of either powder or wire, which is then introduced into the melt pool created through the melting of the substrate using a laser or an electron beam. Process parameters, encompassing scanning speed, wire/powder feed rate, and powder particle size/wire diameter, are subjected to study and optimization in order to achieve the requisite structure and performance of the part. In comparison to the PBF technique, DED permits larger powder sizes, potentially up to three times larger, resulting in heightened molten material flowability. DED finds application in tasks such as the repair of metallic components and coating purposes. Moreover, numerous studies have delved into the realm of utilizing this technique to produce functionally graded materials (FGM) [22].

3.3.4 Binder Jetting

Binder jetting employs a binding agent to adhere to the powdered feedstock, primarily in applications involving metals and polymer materials. The binding agent is dispensed onto a bed of powder, binding the individual powder particles. The build plate descends for the subsequent layer of powder deposition, progressing in this manner until the final part takes shape. The resulting object, also referred to as a green part, is delicate and necessitates subsequent processing – such as heat treatment and high isostatic pressure – to eliminate the binding material and enhance mechanical properties. Numerous factors, including powder characteristics (size, morphology, size distribution), binder material, printing pattern, and layer thickness, exert influence on the attributes of the printed components. Notably, the choice of binder is a pivotal consideration in this process. The binder's printability and low viscosity are critical aspects, as it must possess the capability to deposit material droplets within the powder bed. Notably recognized for its ability to achieve higher production volumes and rates at a reduced cost, binder jetting is widely associated with these advantages [23].

3.3.5 Material Jetting

In the material jetting process, a photosensitive polymer in molten state is solidified through UV light exposure, and it is deposited selectively onto the build plate. Following the completion of each layer, an excess removal step is undertaken using a roller to attain the desired thickness. Subsequently, the printed component is detached from both the build plate and support structures. This technique is particularly effective for creating composite materials that incorporate a blend of soft and hard components, resulting in parts characterized by heightened mechanical attributes.

Material jetting is harnessed for the creation of parts with widespread applications, primarily in modeling and in scenarios not demanding substantial load-bearing capacities. Its applications range from biomedical devices – like surgical training models (especially multicolored bone models), scaffolds for tissue engineering, prostheses, and fixation devices – to the electrical and aviation sectors. Although the use of material jetting in aviation is relatively limited, especially in comparison to other applications, it is in the initial phases, largely restricted to small-scale prototypes. In material jetting, the surface finish can vary between matte and glossy. Notably, process parameters such as layer thickness, surface finish, and build orientation significantly influence the final part's characteristics, encompassing surface roughness, mechanical properties, and dimensional accuracy [24].

3.3.6 Sheet Lamination

The sheet lamination method is divided into two categories, namely laminated object manufacturing (LOM) and ultrasonic additive manufacturing (UAM). In LOM, polymer or metal material sheets are merged through adhesive bonding, thermal bonding, or clamping, and the laser is utilized to cut the part, ultimately achieving the intended design. A similar procedure is employed in UAM, albeit with the distinction that ultrasound is used to fuse the sheets. This technique is primarily recognized for its capability to print intricate geometries, a relatively higher fabrication rate, and cost-effectiveness [25].

3.3.7 Vat Photopolymerization

Vat Photopolymerization, recognized as stereolithography, is the approach employed for photopolymers. As suggested by its name, it involves utilizing a reservoir of photo-curable resin and an ultraviolet laser as the catalyst to solidify the polymer and construct the ultimate part through a layer-by-layer procedure. Upon the full completion of the part's printing, the residual resin in the vat is drained, and the final component is transferred to a UV oven to undergo further curing. Critical process parameters encompass the exposure duration, power of the light source, and the specific wavelength employed. This method facilitates the creation of intricate objects with high-quality surface finish [26].

3.4 Summary and Outlook

The widespread adoption and benefits of 3D printing across diverse applications have spurred researchers to delve deeply into the study and refinement of these methods. Despite the considerable research undertaken in this realm, there remains a need for a thorough understanding of the underlying processing mechanisms to effectively tackle the limitations and obstacles encountered. This entails delving into the intricate dynamics of material transformation, exploring how these changes are affected by the processing conditions, and striving to exert control over the resultant microstructure and properties. Additionally, there is an exploration into the incorporation of novel materials, necessitating comprehensive investigation to harness their potential. In this chapter, we discussed materials and techniques for 3D printing. We elaborated on the properties of materials, i.e., metals, polymers, ceramics, and composites, with a focus on the mostly used materials within each group. Furthermore, different classifications of techniques are addressed, highlighting the main materials associated with each method. The properties, applications, advantages, and drawbacks of the explained materials and techniques are summarized in a table to provide readers with a more distinct understanding and comparison.

References

1 Jasiuk, I., Abueidda, D.W., Kozuch, C. et al. (2018). An overview on additive manufacturing of polymers. *JOM* 70: 275–283.

2 Zhang, J., Song, B., Wei, Q. et al. (2019). A review of selective laser melting of aluminum alloys: processing, microstructure, property and developing trends. *Journal of Materials Science and Technology* 35 (2): 270–284.

3 Astafurov, S. and Astafurova, E. (2021). Phase composition of austenitic stainless steels in additive manufacturing: a review. *Metals* 11 (7): 1052.

4 Herzog, D., Seyda, V., Wycisk, E. et al. (2016). Additive manufacturing of metals. *Acta Materialia* 117: 371–392.

5 Trevisan, F., Calignano, F., Aversa, A. et al. (2018). Additive manufacturing of titanium alloys in the biomedical field: processes, properties and applications. *Journal of Applied Biomaterials & Functional Materials* 16 (2): 57–67.

6 Elahinia, M.H. (2016). *Shape Memory Alloy Actuators: Design, Fabrication, and Experimental Evaluation*. Wiley.

7 Elahinia, M., Moghaddam, N.S., Andani, M.T. et al. (2016). Fabrication of NiTi through additive manufacturing: a review. *Progress in Materials Science* 83: 630–663.

8 Biermann, D., Kahleyss, F., Krebs, E., and Upmeier, T. (2011). A study on micro-machining technology for the machining of NiTi: five-axis

micro-milling and micro deep-hole drilling. *Journal of Materials Engineering and Performance* 20: 745–751.

9 Haberland, C., Meier, H., and Frenzel, J. (2012). On the properties of Ni-rich NiTi shape memory parts produced by selective laser melting. In: *Smart Materials, Adaptive Structures and Intelligent Systems*, vol. 45097. American Society of Mechanical Engineers.

10 Konieczny, B., Szczesio-Wlodarczyk, A., Sokolowski, J. et al. (2020). Challenges of Co–Cr alloy additive manufacturing methods in dentistry—The current state of knowledge (systematic review). *Materials* 13 (16): 3524.

11 Haan, J., Asseln, M., Zivcec, M. et al. (2015). Effect of subsequent hot isostatic pressing on mechanical properties of ASTM F75 alloy produced by selective laser melting. *Powder Metallurgy* 58 (3): 161–165.

12 Tümer, E.H. and Erbil, H.Y. (2021). Extrusion-based 3D printing applications of PLA composites: a review. *Coatings* 11 (4): 390.

13 Kamelian, F.S., Saljoughi, E., Shojaee Nasirabadi, P. et al. (2018). Modifications and research potentials of acrylonitrile/butadiene/styrene (ABS) membranes: a review. *Polymer Composites* 39 (8): 2835–2846.

14 Kumar, R., Singh, R., Ahuja, I.P.S. et al. (2018). Friction welding for the manufacturing of PA6 and ABS structures reinforced with Fe particles. *Composites Part B: Engineering* 132: 244–257.

15 Bahar, A., Belhabib, S., Guessasma, S. et al. (2022). Mechanical and thermal properties of 3D printed polycarbonate. *Energies* 15 (10): 3686.

16 Chen, Z., Li, Z., Li, J. et al. (2019). 3D printing of ceramics: a review. *Journal of the European Ceramic Society* 39 (4): 661–687.

17 Shahrubudin, N., Lee, T.C., and Ramlan, R. (2019). An overview on 3D printing technology: technological, materials, and applications. *Procedia Manufacturing* 35: 1286–1296.

18 Ferreira, R.T.L., Amatte, I.C., Dutra, T.A. et al. (2017). Experimental characterization and micrography of 3D printed PLA and PLA reinforced with short carbon fibers. *Composites Part B: Engineering* 124: 88–100.

19 Wang, X., Jiang, M., Zhou, Z. et al. (2017). 3D printing of polymer matrix composites: a review and prospective. *Composites Part B: Engineering* 110: 442–458.

20 Vanaei, H.R., Shirinbayan, M., Deligant, M. et al. (2021). In-process monitoring of temperature evolution during fused filament fabrication: a journey from numerical to experimental approaches. *Thermo* 1 (3): 332–360.

21 Goh, G.L., Agarwala, S., Goh, G.D. et al. (2018). Additively manufactured multi-material free-form structure with printed electronics. *The International Journal of Advanced Manufacturing Technology* 94: 1309–1316.

22 Haines, M.P., Rielli, V.V., Primig, S. et al. (2022). Powder bed fusion additive manufacturing of Ni-based superalloys: a review of the main microstructural constituents and characterization techniques. *Journal of Materials Science* 57 (30): 14135–14187.

23 Ziaee, M. and Crane, N.B. (2019). Binder jetting: a review of process, materials, and methods. *Additive Manufacturing* 28: 781–801.

24 Tee, Y.L., Tran, P., Leary, M. et al. (2020). 3D Printing of polymer composites with material jetting: mechanical and fractographic analysis. *Additive Manufacturing* 36: 101558.

25 Jadhav, A. and Jadhav, V.S. (2022). A review on 3D printing: an additive manufacturing technology. *Materials Today Proceedings* 62: 2094–2099.

26 Van den Eynde, M. and Van Puyvelde, P. (2018). 3D Printing of poly (lactic acid). *Industrial Applications of Poly (Lactic Acid)* 139–158.

4

Diverse Application of 3D Printing Process

Shohreh Vanaei[1] *and Nader Zirak*[2]

[1] *Department of Bioengineering, Northeastern University, Boston, MA, USA*
[2] *Arts et Métiers Institute of Technology, CNRS, CNAM, PIMM, HESAM University, 75013 Paris, France*

4.1 Introduction

In the landscape of modern innovation, 3D printing has emerged as a transformative technology with a remarkable array of applications across diverse industries. Often referred to as additive manufacturing, 3D printing involves the layer-by-layer construction of three-dimensional objects from digital designs. What began as a niche prototyping technique has rapidly evolved into a versatile tool that is reshaping traditional manufacturing paradigms and pushing the boundaries of creativity and possibility [1].

The inherent flexibility of 3D printing allows for the production of intricate geometries, rapid iteration, and customization that were previously challenging to achieve through conventional manufacturing methods. As a result, the applications of 3D printing span industries as varied as healthcare, aerospace, automotive, fashion, architecture, and more. Its impact is not limited to a single sector; rather, it permeates multiple domains, driving innovation and introducing novel solutions to age-old challenges [2].

This chapter delves into the fascinating world of 3D printing, exploring its wide-ranging applications and the transformative effects it has brought to various industries. From revolutionizing medical procedures through patient-specific implants to enabling the production of lightweight aerospace components, 3D printing has opened up new avenues of creativity, efficiency, and sustainability. As we embark on this exploration, we will uncover the intricate ways in which this technology is reshaping our world and inspiring a future where innovation knows no bounds.

Industrial Strategies and Solutions for 3D Printing: Applications and Optimization, First Edition. Edited by Hamid Reza Vanaei, Sofiane Khelladi, and Abbas Tcharkhtchi.

4.2 3D Printing: Transforming Manufacturing Landscapes

3D printing, also known as additive manufacturing, has emerged as a game-changing technology in the manufacturing domain. Unlike traditional subtractive methods, which involve cutting away material to create a shape, 3D printing adds material layer-by-layer to build objects from the ground up. This revolutionary approach allows for intricate designs, reduced material waste, and the ability to create complex geometries that were previously unattainable. From rapid prototyping to producing end-use parts in various industries, 3D printing has transcended its initial novelty to become a mainstream manufacturing process. At the heart of 3D printing lies a process that intricately constructs objects layer by layer. This methodology starts with a digital 3D model, which is sliced into thin cross-sectional layers. These layers are then sequentially printed using materials ranging from plastics and metals to ceramics and even biological materials. Each layer fuses with the previous one, resulting in a cohesive, three-dimensional object. This layer-by-layer approach not only allows for precise control over design details but also permits the integration of complex internal structures that enhance functionality and performance [3, 4].

The applications of 3D printing span a wide spectrum of industries. In healthcare, it has led to the creation of patient-specific implants, anatomical models for surgical planning, and even the development of bioengineered tissues and organs. In aerospace, 3D printing has facilitated the production of lightweight components with intricate internal cavities, reducing overall aircraft weight and improving fuel efficiency. From fashion and art to automotive and electronics, industries are capitalizing on 3D printing to accelerate innovation, reduce costs, and customize products to individual needs [5].

While 3D printing holds immense promise, it also faces challenges that need to be addressed for its widespread adoption. Material selection, printing speed, surface finish, and post-processing techniques are areas undergoing continuous refinement. Researchers are exploring new materials and optimizing processes to improve the mechanical properties and performance of 3D-printed objects. Moreover, the technology is advancing beyond plastics and metals, venturing into bioprinting for medical applications and printing structures on a larger scale for construction purposes. As 3D printing continues to evolve, it has the potential to redefine industries, business models, and the way we perceive design and manufacturing. With advancements in multimaterial and multicolor printing, the creation of intricate and functional consumer products is becoming increasingly accessible. Additive manufacturing's role in reducing carbon footprint by minimizing waste and enabling localized production is gaining attention. As technology matures and intersects with artificial intelligence and automation, the possibilities for innovation are limitless, heralding a future where customization, sustainability, and efficiency converge. In the upcoming content, endeavors have been made to delve more extensively into the diverse applications of 3D printing across various industries.

4.3 Application of 3D Printing: Different Manufacturing Technology

4.3.1 Fused Deposition Modeling

4.3.1.1 Revolutionizing Prototyping with Fused Deposition Modeling (FDM)

Fused deposition modeling (FDM) stands as a pioneering additive manufacturing technique that has revolutionized rapid prototyping across industries. FDM involves the precise layering of thermoplastic materials, guided by computer-aided designs. This process results in the creation of functional prototypes with intricate details, enabling engineers and designers to visualize and test their concepts quickly. FDM's accessibility and affordability have democratized the prototyping process, allowing for cost-effective iterations, and accelerated product development cycles. From automotive components to medical devices, FDM's impact on prototyping is profound, streamlining the design validation phase and catalyzing innovation [6].

4.3.1.2 Functional End-Use Parts in Manufacturing

Beyond its role in prototyping, FDM has made significant strides in producing functional end-use parts across industries. The advancement of high-performance thermoplastics has paved the way for utilizing FDM to manufacture components that meet stringent mechanical and performance requirements. Aerospace and automotive sectors are harnessing FDM to fabricate lightweight, durable parts with complex geometries. This not only reduces production lead times but also enhances design flexibility, enabling customization and rapid response to changing market demands. FDM's ability to produce end-use parts has reshaped traditional supply chains and manufacturing strategies [7].

4.3.1.3 Medical Advancements Through FDM

The medical field has been profoundly impacted by FDM, offering innovative solutions for patient care and treatment. FDM has enabled the creation of patient-specific surgical guides, implants, and prosthetics with unmatched precision. Surgeons can now plan intricate procedures using anatomically accurate 3D-printed models, improving surgical outcomes, and reducing patient risk. Additionally, FDM's bio-compatible materials have opened doors to bioprinting, where researchers are working toward creating functional tissues and organs for transplantation. FDM's potential to revolutionize personalized healthcare underscores its transformative role in the medical sector [8].

4.3.1.4 Education and Conceptual Learning

FDM has become a cornerstone in educational settings, enriching learning experiences and nurturing creative problem-solving skills. Educational institutions use FDM to bring abstract concepts to life, facilitating hands-on learning across various disciplines. Students can design, prototype, and test their ideas, fostering a deeper understanding of engineering principles. Moreover, FDM's accessibility has empowered students to innovate and develop solutions to real-world challenges, preparing them for the dynamic landscape of modern industry [9, 10].

4.3.1.5 Sustainability and Customization

In an era marked by sustainability concerns, FDM offers a promising path toward environmentally conscious manufacturing. The layer-by-layer approach reduces material waste compared to traditional subtractive methods. Additionally, localized production through FDM minimizes transportation-related carbon emissions. Moreover, FDM's customization capabilities allow manufacturers to produce only what is needed, reducing overproduction and excess inventory. As sustainability becomes integral to business practices, FDM's eco-friendly features make it a valuable tool for responsible production. FDM's versatile applications, from prototyping to end-use parts and from healthcare to education, continue to shape industries and redefine possibilities. Its role as a catalyst for innovation, efficiency, and sustainability underscores its importance in the ever-evolving landscape of additive manufacturing [11].

4.3.2 Stereolithography

4.3.2.1 Precision Prototyping and Beyond with Stereolithography (SLA)

Stereolithography (SLA), a pioneering additive manufacturing technology, has established its presence as a precise and versatile method for prototyping and beyond. SLA employs photopolymerization, where a liquid resin is selectively cured layer-by-layer using ultraviolet light. This meticulous process results in intricately detailed prototypes with exceptionally smooth surface finishes. Its high accuracy and ability to produce complex geometries have made SLA an indispensable tool for industries like automotive, aerospace, and consumer goods, enabling engineers and designers to visualize and evaluate designs with utmost precision [12].

4.3.2.2 Tailoring the Medical Landscape

The medical field has witnessed a transformative impact through the application of SLA. Medical device manufacturers and healthcare professionals rely on SLA to craft patient-specific anatomical models, surgical guides, and implants. These customized solutions are instrumental in enhancing surgical planning, reducing procedure times, and minimizing risks. SLA's ability to replicate patient anatomy with unmatched accuracy has extended to fields such as dentistry and orthopedics, enabling tailored treatments that contribute to improved patient outcomes [13, 14].

4.3.2.3 Architectural and Design Elegance

Architects, artists, and designers have found a canvas for creativity in SLA. SLA offers the ability to translate intricate digital designs into tangible, geometrically complex models and sculptures. This technology allows professionals to experiment with shapes, scales, and textures, enabling a closer connection between the digital and physical worlds. Architects can create detailed scale models of structures, aiding in visualizing designs and conveying concepts to clients and stakeholders. The aesthetic finesse and attention to detail achievable with SLA make it an integral part of architectural visualization and artistic expression [15].

4.3.2.4 Jewelry and Fashion Innovation

The world of jewelry and fashion has embraced SLA for its potential to redefine craftsmanship and design possibilities. SLA enables the creation of intricate jewelry designs that were previously unattainable through traditional methods. From delicate filigree patterns to intricate textures, SLA brings ornate designs to life with exceptional accuracy and surface quality. Fashion designers also utilize SLA to produce unique accessories, accentuating the intersection of technology and style [16].

4.3.2.5 Educational Enrichment and Research

In the realm of education and research, SLA serves as a powerful tool for experiential learning and exploration. Educational institutions leverage SLA to create physical models that aid in understanding complex concepts in fields such as biology, engineering, and architecture. Researchers utilize SLA to fabricate prototypes for testing and analysis, advancing innovations in various domains. SLA's role in facilitating hands-on learning and enhancing research capabilities continues to make it an essential asset in academic and scientific pursuits. SLA's exceptional precision, versatility, and capacity to transform industries and creative fields underscore its significance in the realm of additive manufacturing. Its contributions span from prototyping to personalized healthcare, from architectural visualization to fine art, making it a driving force behind innovation and advancement in diverse sectors [17].

4.3.3 Binder Jetting

4.3.3.1 Redefining Metal Fabrication with Binder Jetting Technology

Binder jetting is a revolutionary additive manufacturing technique that is redefining the landscape of metal fabrication. This process involves selectively depositing a binding agent onto a powder bed, layer-by-layer. The bound regions form the desired part shape, which is then sintered to fuse the particles, resulting in a solid metal component. This technology has made inroads into industries such as automotive and aerospace, where the ability to produce intricate metal parts with high accuracy and speed has transformed traditional manufacturing processes. By offering design freedom and enabling the creation of complex geometries, binder jetting is elevating metal fabrication to new heights of innovation [18].

4.3.3.2 Ceramic Applications and Engineering Advancements

Binder jetting's versatility extends beyond metals into the realm of ceramics. This technique is employed to produce intricate ceramic components with exceptional precision and structural integrity. In industries like electronics and aerospace, ceramic materials play a critical role due to their high-temperature resistance and electrical properties. Binder jetting enables the fabrication of intricate ceramic parts that are used in applications ranging from electrical insulators to turbine blades. The ability to achieve complex designs and exploit the unique properties of ceramics positions binder jetting as a transformative tool in ceramic engineering [19].

4.3.3.3 Transforming Customization and Product Design

Binder jetting's capacity for customization and rapid iteration has made it an invaluable asset in the realm of product design and development. Manufacturers are utilizing this technology to produce prototypes and small-batch production runs that reflect precise design intent. The ability to create intricate internal structures and optimize part geometries allows designers to tailor components to specific functions and applications. This not only accelerates the design process but also facilitates the creation of innovative products that cater to individual customer needs [20].

4.3.3.4 Architectural and Artistic Exploration

Binder jetting's versatility extends even to the realms of architecture and art. Architects and designers leverage this technology to create intricate, full-color models and prototypes that capture their vision with unprecedented precision. These models not only aid in visualizing designs but also serve as functional prototypes for testing structural integrity and spatial arrangements. Moreover, artists are utilizing binder jetting to push the boundaries of creativity, crafting sculptures and artworks that are defined by their intricate details and material diversity [21].

4.3.3.5 Promoting Sustainable Practices and Material Efficiency

In the context of sustainable manufacturing, binder jetting offers notable advantages. The powder-based nature of the process enables material efficiency, as excess powder can be recycled for future use. This reduces material waste compared to traditional subtractive manufacturing methods. Additionally, binder jetting's potential to produce complex shapes with minimal material waste contributes to a more eco-friendly manufacturing approach. As industries increasingly prioritize sustainability, binder jetting's material efficiency aligns well with the principles of responsible production [22].

4.3.4 Power Bed Fusion

4.3.4.1 Empowering Aerospace Innovation with Powder Bed Fusion

Powder Bed Fusion (PBF) stands at the forefront of additive manufacturing technologies, particularly in the aerospace industry. This technique involves layer-by-layer melting of metal powders using high-powered lasers or electron beams to create intricate parts. Its ability to produce complex geometries, consolidate multiple components into a single part, and achieve superior material properties has positioned PBF as a game-changer in aerospace manufacturing. From lightweight structural components to fuel-efficient turbine blades, PBF is propelling the development of advanced aerospace technologies, revolutionizing aircraft performance, and contributing to a more sustainable aviation landscape [23].

4.3.4.2 Medical Advancements Through PBF Techniques

The medical field has witnessed transformative advancements through PBF techniques, particularly in creating patient-specific implants and medical devices. PBF allows for the production of intricate structures with excellent biocompatibility, making it ideal for orthopedic implants, dental restorations, and surgical tools. The precision and control over material properties enable the creation of implants that perfectly match a patient's anatomy, optimizing surgical outcomes and patient recovery. PBF's role in personalized healthcare underscores its potential to revolutionize medical treatments.

4.3.4.3 High-Performance Components in Automotive Engineering

PBF has gained traction in automotive engineering, enabling the production of high-performance components that enhance vehicle efficiency and performance. PBF's ability to create complex lattice structures and optimize part geometries has resulted in lightweight yet durable parts, contributing to reduced fuel consumption and emissions. Moreover, PBF is utilized to manufacture custom components that fit specific vehicle designs, streamlining assembly, and improving overall functionality. As automotive manufacturers seek innovation and sustainability, PBF's contributions are propelling the industry forward.

4.3.4.4 Unlocking Design Possibilities with Customization

The customization potential offered by PBF has broad implications across industries. The ability to create intricate internal structures, lattice networks, and complex geometries opens up design possibilities that were previously unattainable. Jewelry designers leverage PBF to craft one-of-a-kind pieces with intricate details, while industrial designers utilize it to produce functional components with optimized performance. The fusion of design freedom and material precision makes PBF a key enabler of innovation and personalized solutions.

4.3.5 Selective Laser Sintering

4.3.5.1 Elevating Manufacturing Precision with Selective Laser Sintering (SLS)

Selective Laser Sintering (SLS) stands as a cornerstone of additive manufacturing, renowned for its ability to produce intricate and high-precision parts. SLS involves the layer-by-layer fusion of powdered materials, typically polymers or metals, using a high-powered laser. The laser selectively sinters or fuses the powdered particles together, forming a solid component with exceptional accuracy and surface finish. This technology's precision and versatility have led to its widespread application in industries ranging from aerospace and automotive to consumer goods and healthcare [24, 25].

4.3.5.2 Aerospace Innovation Through SLS

Aerospace engineers have harnessed the capabilities of SLS to produce lightweight and durable components that withstand the rigorous demands of flight. From intricate brackets to complex ductwork, SLS-produced parts offer optimized strength-to-weight ratios, leading to increased fuel efficiency and enhanced overall aircraft performance. The ability to create components with minimal material waste and intricate internal structures positions SLS as a vital contributor to the advancement of aerospace technology [26, 27].

4.3.5.3 Medical Devices and Prosthetics

SLS has left a profound impact on the medical field, particularly in the creation of patient-specific medical devices and prosthetics. The customizability and precision of SLS allow for the fabrication of implants that match individual anatomies perfectly, minimizing discomfort and optimizing functionality. Medical professionals leverage SLS to produce intricate surgical tools, orthopedic implants, and even customized hearing aids. The technology's biocompatible materials and design freedom have redefined personalized healthcare solutions.

4.3.5.4 Automotive Engineering and Rapid Prototyping

Automotive manufacturers employ SLS for rapid prototyping and functional testing of vehicle components. SLS-produced prototypes closely resemble the final parts in terms of material properties and surface finish, enabling engineers to assess design integrity and performance early in the development process. The capability to create complex geometries and intricate features contributes to the creation of high-performance automotive components, enhancing vehicle efficiency and safety.

4.3.5.5 Tooling and Manufacturing Efficiency

SLS's role in tooling has streamlined manufacturing processes and reduced lead times. The technology is utilized to produce molds, dies, and jigs with intricate details and complex shapes. SLS-produced tooling components maintain accuracy and durability, allowing for increased production efficiency and quality. Manufacturers leverage the flexibility of SLS to iterate tooling designs quickly and adapt to changing production requirements.

4.3.6 Direct Energy Deposition (DED)

4.3.6.1 Empowering Large-Scale Manufacturing with DED

Direct energy deposition (DED), a versatile additive manufacturing technique, is gaining prominence for its capacity to create large-scale components with speed and precision. DED involves the deposition of material, often in the form of powder or wire, onto a substrate using focused energy sources like lasers or electron beams. This technology's ability to build up parts layer-by-layer, while also allowing for in-process adjustments, is driving its application in industries such as aerospace, oil and gas, and heavy machinery manufacturing [28].

4.3.6.2 Aerospace Advancements with DED

In the aerospace industry, DED is transforming the production of complex components, such as turbine blades and engine parts. DED's rapid material deposition and high deposition rates make it suitable for manufacturing large structural elements with intricate internal features. The technology's ability to repair and refurbish high-value aerospace components also extends their service life, reducing waste and maintenance costs. DED's contribution to the aerospace sector highlights its potential to enhance performance and sustainability [29].

4.3.6.3 Oil and Gas Infrastructure Enhancement

DED's versatility makes it a valuable tool for improving the reliability and longevity of critical infrastructure in the oil and gas sector. This technology is utilized to repair and reinforce equipment exposed to harsh environments, such as offshore platforms and drilling tools. DED's precision and capability to apply material directly onto worn or corroded surfaces restore equipment integrity, reducing downtime and maintenance expenses. Its role in extending the operational life of vital assets showcases DED's importance in critical industries.

4.3.6.4 Tooling and Mold Manufacturing

DED's precision and versatility have found a niche in tooling and mold manufacturing. DED is employed to build intricate and customized tooling components with minimal material waste. This technology is particularly beneficial for creating complex molds for injection molding and die-casting processes. The ability to produce molds directly from digital designs accelerates tooling production and enables rapid iteration of designs, streamlining manufacturing processes, and reducing lead times.

4.3.6.5 Repair and Refurbishment

DED's capability for in situ repair and refurbishment is a game changer in industries reliant on high-value equipment. Sectors such as automotive, aviation, and energy leverage DED to restore worn or damaged components, avoiding the need for costly replacements. This technology's precision allows for localized repair, minimizing the amount of material required and preserving the original component's functionality. DED's role in sustainable maintenance aligns with industries' growing emphasis on efficiency and resource conservation.

4.4 Application of 3D Printing: Industrial Sector

4.4.1 Automotive Innovation Driven by 3D Printing

The automotive industry is currently experiencing a remarkable and dynamic transformation driven by the seamless integration of 3D printing technology. In a bid to enhance the overall efficiency and performance of vehicles, manufacturers within this sector are harnessing the capabilities of 3D printing to craft components that are not only lightweight but also possess exceptional durability. This strategic utilization of 3D printing is leading to a revolution in vehicle construction, where weight reduction and structural integrity are concurrently achieved, resulting in optimized fuel efficiency and overall performance. The traditional process of prototyping within the automotive sector has undergone a significant overhaul due to the incorporation of 3D printing. This technology has streamlined prototyping processes, enabling engineers and designers to iterate designs rapidly and efficiently. The time-intensive cycle of creating and testing new designs has been drastically reduced, allowing for quicker decision-making and more efficient problem-solving. As a result, the industry is witnessing a notable acceleration in the pace of innovation, with designers being empowered to fine-tune their creations in a fraction of the time it once took [30, 31].

Beyond the realms of performance enhancement, 3D printing has unlocked a new realm of possibilities for the automotive sector in terms of customization. Interior elements that were previously constrained by conventional manufacturing methods are now open to unprecedented customization. From personalized dashboard designs to intricately designed air intake systems, the capability to fabricate intricate and unique components has elevated the level of personalization that vehicles can offer to consumers. This level of individuality, once reserved for high-end luxury vehicles, is becoming more accessible across a broader range of automotive products.

As the automotive industry ventures into uncharted territories with novel materials and cutting-edge techniques, 3D printing stands as a cornerstone technology with immense transformative potential. The ability to produce components with diverse materials, intricate geometries, and custom features is setting the stage for a new era in vehicle manufacturing. With 3D printing's inherent agility and adaptability, automotive companies are embracing a future where vehicles are not just a mode of transportation, but also a convergence of innovative design, efficient manufacturing, and unparalleled personalization.

4.4.2 Aerospace Advancements Through 3D Printing

The aerospace industry stands at the forefront of technological advancement, and its adoption of 3D printing is a testament to its commitment to innovation. This transformative technology has permeated every facet of aerospace, ushering in a new era of design and manufacturing methodologies that have far-reaching implications. With its ability to transcend traditional manufacturing limitations, 3D printing is reshaping the industry's landscape by redefining how components are conceptualized, developed, and produced. Central to this shift is the unparalleled flexibility that 3D printing provides concerning both geometry and material utilization. In contrast to customary approaches, which frequently involve removing material from a larger source, 3D printing functions by progressively constructing layers. This grants aerospace engineers the freedom to create and produce intricate components of remarkable precision. An illustrative example is evident in turbine blades, essential for aircraft propulsion, which today exhibit intricately devised structures that enhance aerodynamics and fuel efficiency. This achievement was once unfeasible using conventional manufacturing methods [32].

One of the most significant contributions of 3D printing to aerospace is its ability to reduce material waste. In a sector where lightweight is a priority, the meticulous layering process of 3D printing ensures that material is used only where necessary, minimizing excess and conserving resources. The result is the creation of lightweight structural elements that maintain robustness and reliability, a crucial balance that contributes to improved performance and safety.

Beyond just streamlining the manufacturing process, 3D printing has redefined the pace and efficacy of prototyping. The iterative nature of aerospace design necessitates swift adjustments and frequent revisions. 3D printing's capacity to rapidly produce prototypes allows engineers to visualize and test their concepts in a fraction of the time it once took. This accelerated development cycle enhances decision-making, enabling engineers to refine their designs based on tangible models, thus reducing the likelihood of costly errors in the final production phase.

Moreover, 3D printing has opened the doors to the realm of bespoke components in aerospace. The tailored approach enabled by this technology means that parts can be crafted to suit precise requirements, whether it is optimizing for specific aerodynamics or creating custom interior elements. This bespoke capability not only enhances performance but also supports the realization of groundbreaking innovations that would have been otherwise limited by conventional manufacturing constraints. The aerospace industry's journey with 3D printing is far from over. As engineers continue to experiment with new materials and advanced manufacturing techniques, the integration of 3D printing is set to redefine aerospace production processes and capabilities even further. With its potential to accelerate design cycles, reduce waste, and create custom solutions, 3D printing is poised to remain a cornerstone of aerospace innovation, propelling the industry to new heights of performance, efficiency, and sustainability [33].

4.4.3 3D Printing in Turbomachinery

The application of 3D printing in the domain of turbomachinery engineering heralds a new era of innovation and efficiency. This technology is redefining the manufacturing landscape by allowing the creation of complex, high-performance components with unparalleled precision. From intricately designed blades that optimize aerodynamic performance to lightweight yet robust housings that withstand extreme conditions, 3D printing enables engineers to push the boundaries of design and performance. Moreover, the integration of polymer and polymer composite materials amplifies the capabilities of 3D printing in turbomachinery, introducing enhanced mechanical properties and thermal resistance. This transformative approach expedites the prototyping process, accelerates iterative design cycles, and ultimately leads to the development of more efficient and reliable turbomachinery systems. As industries seek sustainable solutions without compromising on performance, the application of 3D printing in turbomachinery stands as a testament to the technology's pivotal role in shaping the future of engineering excellence.

Recently, FDM and SLS technologies have been used to fabricate different components of the ORC customized system by introducing a micro-turbine-generator-construction-kit (MTG-c-kit). According to the results, while using these technologies can lead to the economical fabrication of different components of this turbomachinery system, lack of suitable fusion is considered to be the cause of air leakage, which is accompanied by reduced efficiency. Fabrication of the impeller of a centrifugal compressor and micro-scale axial turbine has been investigated by SLA technology.

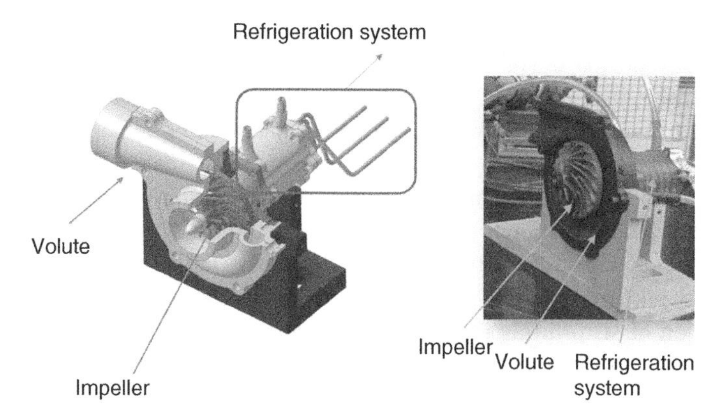

Figure 4.1 An example regarding the part distortion in 3D printing of objects.

The results showed a good ability to use this technology at high rotational speed up to 90,000 rpm. In addition, it was shown to provide a suitable efficiency compared to metal samples at the same conditions. In another study, the prospect of using polymer additive manufacturing by SLA and FDM was investigated to fabricate the different parts of a centrifugal compressor. A high-performance semi-crystalline thermoplastic (polyphenylene sulfide) and a resin based on urethane dimethacrylate have been chosen to fabricate the parts. Figure 4.1 shows a schematic of the centrifugal compressor with different parts consisting impeller, volute, and cooling system. The results showed the suitability of the SLA method compared to FDM to fabricate the rotor at high rotational speeds. The key role of adhesion between polymeric filaments in FDM method was mentioned as a parameter that significantly impacts performance of the compressor.

4.4.4 Food Industry

The food industry is experiencing a culinary revolution driven by the integration of 3D printing technology. This innovative approach is reshaping how food is conceptualized, designed, and produced, offering a new realm of possibilities for both chefs and consumers. 3D printing in the food industry goes beyond mere presentation; it enables the creation of intricate and personalized culinary creations that cater to specific tastes, dietary requirements, and artistic visions. From intricate cake decorations and chocolate sculptures to custom-shaped pasta and intricately designed confections, 3D printing is transcending traditional culinary limitations. This technology empowers chefs to experiment with novel textures, structures, and combinations, resulting in culinary experiences that are as visually stunning as they are flavorful. Moreover, 3D printing also holds the potential for addressing food sustainability and nutritional challenges by customizing ingredients and formulations. As the food industry continues to embrace this cutting-edge technology, 3D printing is poised to redefine the art of gastronomy and inspire a new era of culinary exploration.

4.4.5 Medical Breakthroughs with 3D Printing

The landscape of healthcare has undergone a paradigm shift fueled by the integration of 3D printing technology. This revolutionary advancement has not only redefined medical practices but has also opened up unprecedented avenues for patient-centric care. Through the power of 3D printing, the once unattainable dream of tailor-made implants, prosthetics, and anatomical models has become a reality, ushering in a new era of precision medicine. The implementation of 3D printing has given rise to a realm of possibilities for addressing individual patient needs. Customized implants and prosthetics are now crafted with a level of accuracy and personalization that was previously inconceivable. This breakthrough is not only enhancing patient comfort but is also revolutionizing their quality of life. With the ability to create medical devices that perfectly match anatomical structures, 3D printing is bridging the gap between technological innovation and human well being. Furthermore, the impact of 3D printing extends to the operating room, empowering surgeons with invaluable tools for meticulous planning. Patient-specific anatomical models, crafted through 3D printing, allow surgeons to simulate intricate procedures before they even step into the operating theater. This preoperative insight reduces uncertainties, mitigates risks, and paves the way for more successful outcomes. The amalgamation of 3D printing technology with surgical expertise is not only transforming surgical precision but is also improving patient safety and postoperative recovery [34].

The medical industry's exploration of 3D printing goes beyond devices and models; it ventures into the realm of pharmaceuticals. This technology is being harnessed to innovate drug delivery systems, opening the door to personalized medicine like never before. The concept of tailoring medications to individual patient needs, with dosage and formulation customized, is revolutionizing patient care. This approach has the potential to optimize treatment outcomes while minimizing adverse effects, marking a paradigm shift in how medications are developed and prescribed. In the continuum of progress, 3D printing is rewriting the narrative of patient care and treatment modalities. As the medical community continues to embrace the potential of this technology, the boundaries of possibility are expanding.

From personalized implants to surgical planning and groundbreaking pharmaceuticals, 3D printing is redefining healthcare in ways that were once deemed impossible. It stands as a testament to the synergy between human ingenuity and technological innovation, promising a future where healthcare is not only advanced but truly tailored to the unique needs of each individual.

4.4.6 Electronic Industry

The electronic industry is currently undergoing a profound and transformative shift, catalyzed by the seamless integration of 3D printing technology. This dynamic fusion is not only reshaping established manufacturing methodologies but is also fundamentally altering the entire landscape of how electronic components are conceived, designed, and ultimately fabricated. At the forefront of this evolution stands 3D printing, a technological marvel that has emerged as a veritable game changer in the realm of electronics.

The impact of 3D printing in electronics is profound, evident in its capacity to give rise to intricate and customized electronic devices with unparalleled precision and efficiency. From the intricate interplay of circuitry to the structural intricacies of sensor housings, and even the creation of fully functional prototypes, 3D printing is catalyzing the swift production of highly intricate electronic components that were once marred by complexity when approached through traditional manufacturing channels.

One of the most revolutionary aspects of 3D printing technology in the electronic industry is its ability to offer design flexibility on an unprecedented scale. Engineers and innovators are now empowered to embark on journeys of creative experimentation, where novel shapes, geometries, and configurations can be explored to optimize performance and enhance overall functionality. The electronic components thus created are not only structurally superior but also embody groundbreaking designs that were previously confined by the limitations of conventional manufacturing techniques. As the electronic industry boldly explores the myriad possibilities that 3D printing presents, a transformation of immense magnitude is underway. This transformation promises to redefine the very essence of how electronic devices are manufactured, ushering in an era characterized by streamlined production processes, significantly reduced material waste, and an extraordinary spectrum of design possibilities. With the potential to reshape the industry's foundations, 3D printing is poised to catalyze a new paradigm in electronics – one that embraces versatility, efficiency, and innovation as its guiding principles [35].

4.4.7 Construction Industry: Architecture and Building

The area of architectural design and construction is undergoing a profound renaissance, catalyzed by the relentless evolution of 3D printing technology. This transformative convergence is transcending the conventional boundaries of creativity and enabling architects to manifest their visionary concepts with unprecedented efficiency and precision. Across the spectrum, from intricate scale models to towering full-scale structural elements, architects are harnessing the immense potential of 3D printing to transcend the limitations of traditional construction methodologies. The integration of 3D printing has ushered in a new era where architectural visions are translated into reality with an unmatched level of accuracy. Architects now wield the power to materialize their intricate designs with remarkable fidelity, from the elaborate details of scale models to the grandeur of fully realized structures. This technology serves as a dynamic conduit between creative ideation and tangible fruition, augmenting the architect's capacity to turn abstract concepts into immersive spatial experiences. One of the most remarkable aspects of 3D printing's influence on architecture is its role in prototyping and validation. Through the creation of 3D-printed prototypes, architects can embark on a voyage of exploration into the realm of complex geometries and intricate design decisions. This pre-construction visualization aids in unraveling the intricacies of design choices, offering insights into spatial dynamics, lighting effects, and overall aesthetic appeal. By providing a tangible and three-dimensional preview, 3D printing empowers architects to refine their concepts before construction, reducing uncertainties, and enhancing the precision of their designs.

Sustainability lies at the heart of modern architectural discourse, and here, 3D printing has a transformative role to play. Forward-thinking architects are investigating the application of 3D printing to produce eco-friendly materials and minimize waste in construction processes. This approach not only reduces the environmental impact but also aligns with the ethos of sustainable development. By optimizing material usage and exploring renewable resources, 3D printing contributes to the creation of structures that embody both aesthetic innovation and ecological responsibility. As architects continue to unravel the myriad possibilities embedded within 3D printing, the built environment stands on the cusp of an era marked by ingenuity and boundless creativity. The convergence of technology and design is revolutionizing architectural processes, transcending the conventional norms of construction, and nurturing a fertile ground for innovation. With each layer that 3D printing adds to the landscape of architecture, the industry's capacity for innovation and aesthetic expression expands, heralding a future where the physical world mirrors the limitless imagination of the human mind.

4.4.8 Fashion Industry

The fashion industry is undergoing a profound metamorphosis with the infusion of 3D printing technology. This creative convergence is transcending traditional design methodologies and revolutionizing the way garments and accessories are conceptualized, crafted, and showcased. 3D printing has emerged as a catalyst for innovation, allowing designers to push the boundaries of form, texture, and structure in ways previously unimaginable. From avant-garde couture to everyday wear, designers are leveraging the dynamic capabilities of 3D printing to fuse artistry with technology, giving rise to a new era of fashion aesthetics. The integration of 3D printing into the fashion world signifies a shift toward a realm of design that is not bound by conventional limitations. Designers now possess the means to translate intricate digital designs into tangible fashion pieces with impeccable precision. This transformative technology empowers designers to experiment with novel shapes, textures, and architectural forms, creating garments that are as sculptural as they are wearable. The fusion of creativity and technology transcends the confines of traditional textiles, paving the way for garments that are true works of art.

The impact of 3D printing extends beyond the design studio and runway; it ushers in a paradigm shift in the relationship between fashion and individuality. Customization takes center stage as 3D printing allows for tailored fashion experiences, where each piece can be personalized to fit the wearer's unique preferences and measurements. This level of bespoke craftsmanship transforms fashion into an intimate expression of identity, where consumers become co-creators in the design process. Furthermore, 3D printing offers a sustainable alternative within the fashion industry. By minimizing material waste through precise additive manufacturing, this technology aligns with the growing demand for eco-conscious production methods. Designers are exploring recycled and bio-based materials, revolutionizing the fashion supply chain, and fostering a more sustainable future for the industry. As the fashion industry continues to explore the myriad possibilities of 3D printing, a new realm of creativity and innovation is being unveiled. The synthesis of technology and design has paved the way for garments that are as functional as they are aesthetically striking. By reshaping the way fashion is envisioned, created, and experienced, 3D printing is not merely changing the fabric of fashion – it is sewing the threads of a transformative future where creativity knows no bounds.

Table 4.1 A summary of different AM methods with respect to their applications.

Method	Application	Key highlights
FDM	Automotive Aerospace Turbomachinery Food Medical Electronics Fashion	Prototyping, custom tooling Lightweight components in aerospace Complex geometries in turbomachinery Culinary tools and molds Medical prosthetics and implants Custom electronics housings Construction prototypes Fashion accessories
SLA	Automotive Aerospace Turbomachinery Food Medical Electronics Construction Fashion	Intricate prototypes Wind tunnel models in aerospace Optimized designs in turbomachinery Custom edible creations Surgical guides and dental models Detailed casings for electronics Architectural models Bespoke jewelry
SLS	Automotive Aerospace Turbomachinery Food Medical Electronics Construction Fashion	Functional prototypes and end-use parts Engine components in aerospace High-temperature applications in turbomachinery Edible structures Custom implants and prosthetics Custom electronics housings Construction prototypes and molds Wearable tech and footwear
DED	Automotive Aerospace Turbomachinery Food Medical Electronics Construction Fashion	Metal tooling and functional parts Engine components and repairs in aerospace Repair and optimization of turbine parts 3d printed food structures Custom implants and prosthetics Metal additive manufacturing for electronics Construction components Custom jewelry and accessories

4.5 Summary

The application of 3D printing has ushered in a new era of manufacturing and creativity, revolutionizing industries and shaping the way we design, create, and interact with objects. Its versatility spans across sectors as diverse as healthcare, aerospace, art, consumer goods, and more. The technology's ability to rapidly prototype, produce complex geometries, and customize designs has led to faster innovation cycles, reduced material waste, and increased cost-effectiveness. As 3D printing continues to evolve and researchers explore new materials and methods, its potential for disruption and innovation only grows stronger. Whether it is in healthcare, where it is redefining personalized medicine, or in manufacturing, where it is optimizing supply chains, 3D printing stands as a testament to the power of combining digital ingenuity with physical creation, driving us toward a future where imagination knows no bounds. The different AM techniques with respect to their applications are summarized in Table 4.1.

References

1. Perkins, I. and Skitmore, M. (2015). Three-dimensional printing in the construction industry: a review. *International Journal of Construction Management* 15 (1): 1–9.
2. Lee, J.-Y., An, J., and Chua, C.K. (2017). Fundamentals and applications of 3D printing for novel materials. *Applied Materials Today* 7: 120–133.
3. Campbell, T., Williams, C., Ivanova, O. et al. (2011). Could 3D printing change the world. *Technologies, Potential, and Implications of Additive Manufacturing, Atlantic Council, Washington, DC* 3: 1–16.
4. Babbar, A., Rai, A., and Sharma, A. (2021). Latest trend in building construction: three-dimensional printing. In: *Journal of Physics: Conference Series*. IOP Publishing.
5. Ludwig, T., Stickel, O., Boden, A. et al. (2014). Towards sociable technologies: an empirical study on designing appropriation infrastructures for 3D printing. In: *Proceedings of the 2014 conference on Designing interactive systems*.
6. Rahim, T.N.A.T., Abdullah, A.M., and Md Akil, H. (2019). Recent developments in fused deposition modeling-based 3D printing of polymers and their composites. *Polymer Reviews* 59 (4): 589–624.
7. Hassani, V. (2020). An investigation of additive manufacturing technologies for development of end-use components: a case study. *International Journal of Pressure Vessels and Piping* 187: 104171.

8 Awad, A., Gaisford, S., and Basit, A.W. (2018). Fused deposition modelling: advances in engineering and medicine. *3D Printing of Pharmaceuticals* 107–132.

9 Dilling, F. and Witzke, I. (2020). The use of 3D-printing technology in calculus education: concept formation processes of the concept of derivative with printed graphs of functions. *Digital Experiences in Mathematics Education* 6: 320–339.

10 Ford, S. and Minshall, T. (2019). Invited review article: where and how 3D printing is used in teaching and education. *Additive Manufacturing* 25: 131–150.

11 Gebler, M., Uiterkamp, A.J.S., and Visser, C. (2014). A global sustainability perspective on 3D printing technologies. *Energy Policy* 74: 158–167.

12 Mukhtarkhanov, M., Perveen, A., and Talamona, D. (2020). Application of stereolithography based 3D printing technology in investment casting. *Micromachines* 11 (10): 946.

13 Douroumis, D. (2019). 3D printing of pharmaceutical and medical applications: a new era. *Pharmaceutical Research* 36 (3): 42.

14 Hsiao, W.-K., Lorber, B., Reitsamer, H. et al. (2018). 3D printing of oral drugs: a new reality or hype? *Expert Opinion on Drug Delivery* 15 (1): 1–4.

15 Grigoryan, B., Sazer, D.W., Avila, A. et al. (2021). Development, characterization, and applications of multi-material stereolithography bioprinting. *Scientific Reports* 11 (1): 3171.

16 Pasricha, A. and Greeninger, R. (2018). Exploration of 3D printing to create zero-waste sustainable fashion notions and jewelry. *Fashion and Textiles* 5 (1): 1–18.

17 Pinger, C.W., Geiger, M.K., and Spence, D.M. (2019). Applications of 3D-printing for improving chemistry education. *Journal of Chemical Education* 97 (1): 112–117.

18 Dilberoglu, U.M., Gharehpapagh, B., Yaman, U. et al. (2017). The role of additive manufacturing in the era of industry 4.0. *Procedia Manufacturing* 11: 545–554.

19 Peng, E., Zhang, D., and Ding, J. (2018). Ceramic robocasting: recent achievements, potential, and future developments. *Advanced Materials* 30 (47): 1802404.

20 Ragelle, H., Rahimian, S., Guzzi, E.A. et al. (2021). Additive manufacturing in drug delivery: Innovative drug product design and opportunities for industrial application. *Advanced Drug Delivery Reviews* 178: 113990.

21 Vanderploeg, A., Lee, S.-E., and Mamp, M. (2017). The application of 3D printing technology in the fashion industry. *International Journal of Fashion Design, Technology and Education* 10 (2): 170–179.

22 Lores, A., Azurmendi, N., Agote, I. et al. (2019). A review on recent developments in binder jetting metal additive manufacturing: materials and process characteristics. *Powder Metallurgy* 62 (5): 267–296.

23 Dzogbewu, T.C. (2020). Laser powder bed fusion of Ti6Al4V lattice structures and their applications. *Journal Of Metals, Materials and Minerals* 30 (4): 68–78.

24 Mazzoli, A. (2013). Selective laser sintering in biomedical engineering. *Medical & Biological Engineering & Computing* 51: 245–256.

25 Korkmaz, M.E., Gupta, M.K., Robak, G. et al. (2022). Development of lattice structure with selective laser melting process: a state of the art on properties, future trends and challenges. *Journal of Manufacturing Processes* 81: 1040–1063.

26 Vashishtha, V.K., Makade, R., and Mehla, N. (2011). Advancement of rapid prototyping in aerospace industry-a review. *International Journal of Engineering, Science and Technology* 3 (3): 2486–2493.

27 Jing, L.L., wei Cheng, K., Wang, F. et al. (2020). Application of selective laser melting technology based on titanium alloy in aerospace products. In: *IOP Conference Series: Materials Science and Engineering*. IOP Publishing.

28 Gibson, I., Rosen, D., Stucker, B. et al. (2015). Directed energy deposition processes. In: *Additive Manufacturing Technologies: 3D Printing, Rapid Prototyping, and Direct Digital Manufacturing*, Springer, 245–268.

29 Zeng, Q., Xu, Z., Tian, Y. et al. (2016). Advancement in additive manufacturing & numerical modelling considerations of direct energy deposition process. In: *Proceeding of the 14th International Conference on Manufacturing Research: Advances in Manufacturing Technology XXX* (ed. Y.M. Goy and K. Case). Amsterdam: IOS Press.

30 Lecklider, T. (2017). 3D printing drives automotive innovation. *EE-Evaluation Engineering* 56 (1): 16–20.

31 Lim, C.W.J., Le, K.Q., Lu, Q. et al. (2016). An overview of 3-D printing in manufacturing, aerospace, and automotive industries. *IEEE Potentials* 35 (4): 18–22.

32 Wimpenny, D.I., Pandey, P.M., and Kumar, L.J. (2017). *Advances in 3D Printing & Additive Manufacturing Technologies*, vol. 1. Springer.

33 Chapiro, M. (2016). Current achievements and future outlook for composites in 3D printing. *Reinforced Plastics* 60 (6): 372–375.

34 Pavan Kalyan, B. and Kumar, L. (2022). 3D printing: applications in tissue engineering, medical devices, and drug delivery. *AAPS PharmSciTech* 23 (4): 92.

35 Shahrubudin, N., Lee, T.C., and Ramlan, R. (2019). An overview on 3D printing technology: technological, materials, and applications. *Procedia Manufacturing* 35: 1286–1296.

5

Redefining Fabrication: Emerging Challenges in the Evaluation of 3D-printed Parts

Xiaofan Luo[1], Mengxue Yan[2], Kaddour Raissi[3], and Amrid Mammeri[3,4]

[1] *Polymaker LLC, Houston, TX, USA*
[2] *School of Mechanical Engineering, Changshu Institute of Technology, Changshu, Suzhou, Jiangsu, China*
[3] *Arts et Métiers Institute of Technology, CNAM, LIFSE, HESAM University, 75013, Paris, France*
[4] *Valeo Thermal Systems, La verriere-Paris, France*

5.1 Introduction: Scope and Definition

Over the course of history, the concept of "3D printing" has undergone a transformation in its definition. In contemporary times, the term "3D printing" is most commonly used interchangeably with "additive manufacturing (AM)." This refers to a category of digital manufacturing techniques wherein physical objects are constructed by systematically adding materials in a layer-by-layer manner. Throughout the following discourse, we will adhere to this prevalent interpretation.

According to the standards set by American Society for Testing and Materials (ASTM), the majority of existing 3D printing technologies can be classified into seven distinct process categories. These categories are material extrusion, material jetting, binder jetting, vat photopolymerization, sheet lamination, powder bed fusion, and directed energy deposition. Among these, material-extrusion-based 3D printing, often abbreviated as ME-3DP, encompasses a range of technologies that employ extrusion mechanisms to position materials within a three-dimensional space. Due to its unparalleled popularity – more on the reasons behind this popularity will be discussed later – and significant potential, ME-3DP will be the primary focal point in this chapter. The exploration of this chapter in the book will delve into the present state, existing challenges, and future prospects of this highly promising technology. In essence, it explains how the meaning of "3D printing" has changed over time to align with AM, the

Industrial Strategies and Solutions for 3D Printing: Applications and Optimization, First Edition. Edited by Hamid Reza Vanaei, Sofiane Khelladi, and Abbas Tcharkhtchi.

seven categories into which ASTM classifies 3D printing technologies, and the decision to concentrate primarily on ME-3DP due to its widespread usage and promising prospects.

5.2 Historical Review

ME-3DP was among the earliest 3D printing technologies invented, with the first patent appearing in the late 80s [1]. This led to the founding of Stratasys, which, over the years, grew to become (and still remains) one of the largest and most impactful 3D printing companies in the world. In the first two decades of ME-3DP, the market was dominated largely by Stratasys with its fused deposition modeling or FDM® machines. These were fully integrated machines used primarily in rapid prototyping (similar to other 3D printing technologies at the time) applications.

There was a significant shift in the technological and market landscape of ME-3DP that started in the late 2000s. This shift was driven primarily by two landmark events – (i) the open-source movement known as the RepRap (which stands for "replicating rapid prototype") project, and (ii) expiration of the original Stratasys patents. The market soon saw an explosion of players (most of which were hardware startups) developing RepRap-inspired desktop 3D printers that were mechanistically similar (if not identical) to the Stratasys machines [2]. Some of these companies grew aggressively either to become major market players themselves or to be acquired by incumbent market leaders. The net result was an emerging market with increasingly diversified products. For example, today there are hundreds, if not more, FFF (which stands for fused filament fabrication and is becoming a more standard term than FDM) printers in the market, with prices ranging from a few hundred dollars all the way to hundreds of thousands of dollars. Another major shift was the change from fully closed product architectures (as was the case in all early-era 3D printing companies) to an open ecosystem composed of specialized providers of machines, materials, components, and software. This shift significantly accelerated the democratization and innovation in ME-3DP and allowed the technology to penetrate into markets that were not accessible previously due to various barriers. Today, ME-3DP is the most popular 3D printing technology measured by both machine shipment (over 1 million per year globally) and user base (also in millions).

While the majority of ME-3DP printers today are FFF printers that still work more or less the same way as the original Stratasys machines, there have been process innovations over the years that have greatly expanded and enriched the technological landscape of 3D printing process (Figure 5.1).

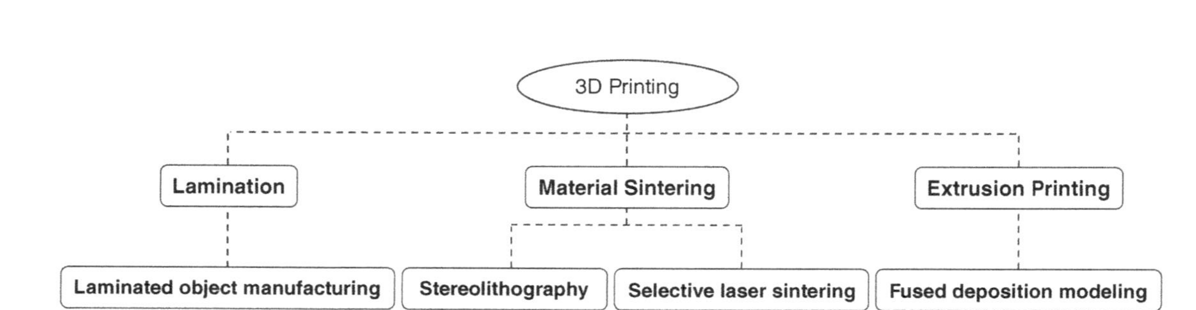

Figure 5.1 Process variations in 3D printing.

One of the most notable examples is the use of ME-3DP to produce continuous fiber-reinforced composite structures, first introduced by the US-based Markforged and later becoming a hot topic both industrially and academically. This technology can be used to produce structures with very high stiffness and strength, allowing ME-3DP to enter new property spaces that were not previously possible.

Another major process innovation is the invention of large-scale, pellet-based ME-3DP systems. The most notable work was the big area additive manufacturing (BAAM) system developed by the Oak Ridge National Lab (ORNL). This has expanded ME-3DP in another dimension – size. Today, ME-3DP is the best 3D printing technology for producing large polymer- and composite material-based single structures, which can go up to several meters in lateral dimensions. Unlike in most application scenarios where 3D printing is explored as an alternative to traditional manufacturing technologies, large-scale pellet-based ME-3DP is opening entirely new territories, i.e. there is simply no other (at least economically feasible) way of producing these large structures except BAAM. This technology is already generating transformative impact in areas such as architecture and composite tooling.

Direct ink writing (DIW) represents another ME-3DP process with some unique distinctions. Instead of applying heat to process thermoplastics (as in the case of all other ME-3DP processes), DIW uses liquid "inks" with properly designed rheological properties and curing (usually thermosetting) chemistries to produce objects, thus opening doors to new materials and functionalities. Although the commercial and industrial use of DIW is currently limited, we do see great future promises of this technology in many areas.

The development in materials for ME-3DP is also noteworthy. The shift to much more open systems encouraged new researchers and companies to join the material development effort for ME-3DP. Suddenly, the variety of materials available for ME-3DP grew substantially, to the point that today it is hard to find a thermoplastic material that has not been at least explored for ME-3DP. This move is greatly favored by customers as they are no longer "locked in" by the machine manufacturer and can have more choices and flexibility in their applications.

The rapid growth and various technological achievements in ME-3DP in the last decade are definitely something that shall be praised and celebrated. However, we should see that the majority of these efforts have been to expand the breadth of the technology, e.g. applying the same principle at different scales (from millimeters to meters), developing different process variations, and exploring different materials. In contrast, we have not seen the same level of progress in the depth of the technology, e.g. deeper understanding of the complex physics involved in ME-3DP and the development of next-generation tools to improve the performance of the technology. As a result, many of the fundamental challenges in ME-3DP (which will be analyzed further in the rest of the chapter) remain largely unsolved and still haunt the technology today. Yes, users may have more choices, but they are not necessarily better choices. Most of the commercially available systems still suffer from poor reliability, low speed, and unpredictable properties among others, that pale in comparison to traditional manufacturing processes such as injection molding.

It is widely anticipated that AM will play a much larger role in the manufacturing landscape. AM is also a vital technology in the global fight against climate change, with the potential to reduce up to 27% of the global energy consumption according to one study [3]. However, as the most popular AM process today, ME-3DP still has a long way to go before it can claim to be a modern, mainstream manufacturing technology.

In what follows, we will dive deeper into the fundamental technological challenges faced by ME-3DP, why the challenges exist, and roadmaps of potential solutions.

5.3 Technological Challenges in ME-3DP

In the examination of challenges within the context of material-extrusion-based 3D printing (ME-3DP), it is required to adopt a systematic approach that involves differentiating between *symptoms* and *root causes*. This methodology has been chosen with the intention of addressing prevalent misunderstandings and misconceptions surrounding ME-3DP. By employing this approach, the goal is to offer a transparent, accurate, and ideally unbiased portrayal of the issues and hurdles that are encountered within this field. By distinguishing between symptoms and root causes, the aim is to pinpoint the surface-level manifestations of challenges (symptoms) and delve deeper into the underlying fundamental factors that give rise to these difficulties (root causes). This differentiation will enable researchers/industries to present a more comprehensive and insightful analysis of the challenges faced in ME-3DP [4].

In doing so, we aspire to untangle any confusion or inaccuracies that might have arisen due to incomplete understanding or misinterpretation of the problems at hand. The outcome of our endeavor is to provide a coherent and precise assessment of the challenges in ME-3DP, fostering a better comprehension of the field and paving the way for more effective solutions and advancements.

5.3.1 The Symptoms of ME-3DP

The term "symptoms" primarily alludes to the tangible aspects that individuals directly engaged in the practical application of ME-3DP encounter and perceive within real-world scenarios. Accurately comprehending these symptoms carries significant practical importance, as it serves not only as a practical consideration but also as an essential initial stride toward identifying the genuine issues and ultimately steering the formulation of effective solutions. These symptoms, drawn from the authors' extensive experience spanning over a decade in the industry and their observance of its evolution, encapsulate the essence of what users face when working with ME-3DP. However, it is important to note that this compilation of symptoms is not intended to be exhaustive; rather, it aims to spotlight what the authors consider to be the fundamental, distinguishing, and persistent indicators in contemporary ME-3DP.

These identified symptoms – poor process reliability, low printing speed, part distortion, and unpredictable material properties – transcend mere theoretical conjecture; they stem from the real-world intricacies of ME-3DP operations and applications. Each symptom represents a concrete issue that practitioners and professionals encounter during their interactions with the technology. These symptoms collectively form a crucial foundation for understanding the existing challenges inherent to ME-3DP. It is imperative to acknowledge that these symptoms are not just random observations, but rather the result of a meticulous analysis grounded in practical experience. They collectively serve as a reference point for establishing a baseline understanding of the key issues and hurdles associated with ME-3DP. By grasping these symptoms, the groundwork is laid for tackling the deep-rooted problems and advancing the development of targeted solutions that can enhance the performance, reliability, and potential of ME-3DP processes.

5.3.1.1 Poor Process Reliability

Good reliability is a basic requirement for any modern industrial manufacturing process. Achieving this, however, turns out to be not so easy for ME-3DP. The poor process reliability can manifest itself in different ways – failed prints, printing defects, and inconsistent quality, just to name a few, and they occur in pretty much all ME-3DP systems ranging from entry-level desktop printers to expensive "industrial" machines. Generally speaking, these reliability issues can fall into two categories: those that are equipment-related and those that are process-related. Equipment-related issues have become rarer over the years as improvements in equipment are more straightforward. For example, you are much less likely to get a purely mechanical issue on today's ME-3DP machines as the motion systems have become considerably more robust. On the other hand, process-related reliability issues are more complex and much harder to tackle, and this is worth examining further.

It is required to explore the concept of "intended function," as defined by ASQ in their description of reliability, within the context of ME-3DP. This exploration is crucial as it shapes the way we perceive the reliability or lack thereof in the process. In the realm of injection molding, a part is created using a predetermined mold, essentially ensuring that the mold produces a singular part according to its cavity design. Conversely, the paradigm differs in ME-3DP, where the system is expected to manufacture a wide array of parts, provided they adhere to specific geometric criteria, such as maximum dimensions and feature sizes. Furthermore, much like injection molding, contemporary ME-3DP systems are engineered to accommodate various materials. The precise range of supported materials often lacks a precise definition, yet the fundamental principle remains – prospective buyers demand ME-3DP machines capable of working with a diverse range of materials rather than being restricted to a single fixed material. Consequently, the primary intended function of the majority of ME-3DP systems is 2-fold: (i) the capacity to craft parts of diverse geometries within specific limitations, and (ii) the ability to execute printing using multiple materials. Naturally, the execution of this intended function must be consistent and with minimal instances of failure.

In a typical ME-3DP workflow, a digital file (in STL or, more recently, 3MF format) is processed in a "slicer" program that generates process commands and tool paths based on a fixed set of user-defined parameters. This parameter set, which contains a large number (tens to hundreds) of process variables such as nozzle temperature, infill density, layer height, and various printing speeds, is often regarded as a "print profile" and is vital in determining the outcome of printing. The critical question is how "universal" a print profile can be, i.e. whether we can have a "perfect" print profile that works 100% of the time regardless of part geometries and even materials. Unfortunately, the answer is no (we will show why in later sections), at least not with today's software tools. Some basic print profiles are often provided by the system provider, but they are meant more as a starting point and users are expected to make adjustments based on different part geometries and materials. There are some general guidelines on how to adjust the print profile, but they are far from comprehensive. In reality, users often have to rely heavily on their personal experience together with lengthy trial-and-error, as it is almost impossible to guarantee that a print profile will work without actually printing the part, especially for more challenging geometries and materials. It is common for a user to have a number of failed prints before a successful one. This is often perceived as the system being "unreliable." Even if one takes the tedious work to finally obtain a "good" print profile (i.e. one that works in a particular case), this success is difficult to transfer among different printers, materials, or, most importantly, parts of different geometries. In other words, one often has to go through the same experience-based, trial-and-error process to when a new printer, material, and/or part is involved. This further strengthens people's perception that the system is "unreliable."

Of course, those process-related reliability issues can be mitigated by limiting the part geometries and material options, and this is indeed a practice adopted by most system providers today. In the extreme case, if we restrict the system to only print one material and one part, this issue will completely disappear. But this is not what the intended function is for ME-3DP, and certainly not the future of the technology.

5.3.1.2 Low Printing Speed

Individuals who have experience with the present-day ME-3DP will perceive it as a notably sluggish procedure. The fabrication of a moderately sized component can demand several hours, while more substantial parts might necessitate several days. This stands in stark contradistinction to injection molding, where the time taken for each cycle is gauged in seconds and minutes. It is unsurprising that the constrained printing speed is universally recognized as the primary obstacle preventing the broader acceptance of ME-3DP.

Why is the speed of printing important? Granted, there are certain scenarios where the duration of printing is crucial – for instance, if the printing process of a surgical guide extends for too long, there is a risk of missing the optimal surgical window. However, of greater significance, slower printing speeds directly result in elevated production expenses. Therefore, the underlying inquiry driving the pursuit of speed is whether we can reduce manufacturing costs. It appears that enhancing printing speed is an effective approach toward this goal [5].

An alternative strategy for decreasing manufacturing expenses involves the simple act of extruding more materials at an accelerated pace. For instance, when considering a fixed printing speed (a parameter defined in ME-3DP as the travel velocity of the printhead in millimeters per second), if we manage to augment the quantity of extruded material, we can subsequently attain a heightened throughput (measured in cubic millimeters per second or kilograms per hour). It is worth noting that throughput, rather than speed, directly governs the costs involved. This strategy commonly entails employing more robust extruders equipped with larger nozzles, along with printing at increased layer heights or widths. The most extreme illustration of this approach is the BAAM process, which employs screw extruders and can accomplish throughputs reaching approximately 50 kilograms per hour. Nevertheless, the drawback of this tactic is quite evident – as nozzle sizes and flow rates increase, the spatial resolution diminishes, which clarifies why this method has been confined to producing larger, geometrically straightforward components lacking intricate details.

Hence, elevating the print speed appears to be the subsequent logical step and fundamentally the most effective means to reduce manufacturing expenses without making concessions. However, it has also emerged as the most intricate technological hurdle (which will be elaborated upon in the forthcoming section). Numerous companies have endeavored to address this challenge, and although specific domains have witnessed advancements and innovative solutions, there has not been an overarching enhancement in print speed over the past decade. Presently, the majority of FFF printers operate within the range of 20 to 120 millimeters per second (mm/s), figures that are quite comparable to those of printers crafted a decade ago. It is only quite recently that a few new "high-speed" printers have begun to emerge, yet their efficacy is still pending evaluation by the market.

5.3.1.3 Part Distortion

Another prevalent problem in ME-3DP is part distortion, which involves significant deviations in the physical measurements and geometric characteristics of the printed object from its digital design. This issue can result in print failures and the production of nonfunctional components.

Warpage is the most frequently observed type of part distortion in ME-3DP. It typically initiates at the base where the part makes contact with the build plate and worsens progressively during the printing process. The extent of warpage occurrence varies across different printers, materials, and part shapes. Some general tendencies include semi-crystalline polymers displaying more warping compared to amorphous ones, amorphous polymers with higher glass transition temperatures (T_g) being more susceptible to warping than those with lower T_g values, polymers containing high aspect-ratio fillers like chopped carbon fibers generally exhibiting reduced warping, and larger parts being more prone to warping than smaller counterparts.

While adjusting printing settings can at times alleviate this issue, solutions are not always straightforward. Some remedies are well established and widely accepted – for example, elevating the build envelope temperature has proven effective in reducing warping for high-T_g amorphous polymers. However, in many instances, the connection between printing profiles and warpages remains unclear and difficult to predict. The precise mechanism behind warpage is not fully comprehended, especially concerning the origins and development of residual stresses that trigger warping. What is evident, though, is that warpage is a multifaceted process influenced by numerous contributing factors and physical phenomena.

Apart from warpage, various other types of part distortion exist. Although their sources differ, they all exhibit comparable levels of intricacy akin to warpage. These distortions are challenging to predict and manage in real-world scenarios, and their mitigation proves to be an even more formidable task.

5.3.1.4 Unpredictable Properties

The final major characteristic found in contemporary ME-3DP is the substantial and unpredictable alterations in the attributes of printed components, even when using the same materials. To exemplify this, let us begin by examining how we forecast the properties of conventionally manufactured parts. In many instances, we do not directly measure the part; instead, we measure the material for its inherent properties of interest – factors like Young's modulus, ultimate strength, conductivity, flame retardance, and so on. The underlying assumption is that the part will inherit all these material properties, regardless of the manufacturing method employed. While this assumption holds true for most traditional manufacturing processes such as injection molding and CNC machining, it does not apply to ME-3DP. In this AM method, both the materials used and the process itself can profoundly impact the properties of the resulting part. As a distinguished researcher in the field of ME-3DP once succinctly stated, "The central challenge of 3D printing is that you cannot separately control the shape and properties." This statement precisely encapsulates the essence of ME-3DP.

While there has been some comprehension regarding how the manufacturing process influences the characteristics of parts (a topic we will explore later), the intricate interplay of numerous factors renders the ability to foresee and manage outcomes in reality challenging, if not unachievable. Even when employing the "same" printer and print profile, seemingly arbitrary discrepancies in properties can manifest from one part to another.

Furthermore, not all properties react to the process uniformly. Adding to the complexity, distinct regions, or voxels within a single part can be subject to differing effects, resulting in varied properties – illustrating a heterogeneous distribution of attributes within the same component (which extends beyond mere anisotropy). Some might argue that the fundamental issue is the comparatively reduced strength of parts produced via ME-3DP compared to those crafted through conventional manufacturing techniques like injection molding. While acknowledging this general observation, especially in the context of mechanical properties, it is posited that the true challenge is less about the inherent property weaknesses and more about their unpredictability. It is well understood that the ultimate tensile strength of a part along the z-axis is only a fraction of that along the X- and Y-axes. However, as long as the recognized strength meets the application's requirements, the potential z-axis weakness is not inherently problematic, especially from a pragmatic (rather than purely scientific) standpoint. The issue arises when strength variations occur across different parts, with no method for anticipation or control. How can we determine whether a specific part satisfies the strength criteria or not? In many instances, testing each individual part, particularly for mechanical strength (which involves destructively breaking the part), is not feasible. Just like the other mentioned issues, this difficulty emerges due to the anticipation that ME-3DP can generate various parts with diverse shapes and materials, all while imposing minimal limitations. If we narrow down the range of permissible geometries and materials, this challenge could become less pronounced. As observed by the authors, this continues to be the primary approach adopted within the industry today.

5.3.2 The Root Cause

5.3.2.1 Process Complexity: ME-3DP vs Injection Molding

The process complexity between ME-3DP and injection molding is marked by distinct differences stemming from their underlying methodologies. ME-3DP involves the incremental layering of material to create a three-dimensional object directly from a digital design. This layer-by-layer approach allows for intricate geometries and customization, as well as the potential to consolidate complex assemblies into a single printed part. However, the additive nature of the process introduces challenges related to material adhesion, support structure generation, and post-processing steps to achieve the desired surface finish and mechanical properties. Furthermore, the interplay between layer adhesion, cooling rates, and thermal stresses during printing can lead to part warping and anisotropic properties, necessitating careful optimization and consideration of print parameters.

On the other hand, injection molding, a subtractive manufacturing method, involves injecting molten material into a mold cavity to form the desired shape. The high pressures and temperatures used in injection molding enable the production of parts with excellent surface finish, dimensional accuracy, and reproducibility. The process is particularly suited for high-volume production due to its rapid cycle times and consistent outcomes. However, injection molding requires the creation of molds, which can be costly and time-consuming, making it less agile for rapid design iteration or producing small quantities of parts. Additionally, intricate geometries and internal cavities can pose challenges in mold design and material flow, potentially leading to defects like air traps or incomplete filling.

While ME-3DP offers versatility and reduces tooling costs, it often grapples with issues of material compatibility, layer resolution, and post-processing complexity. Injection molding excels in producing high-quality parts at scale but is constrained by the mold creation process and may struggle with complex geometries. Both processes represent valuable tools within the manufacturing landscape, each suited to specific contexts and requirements, highlighting the nuanced interplay between complexity, adaptability, and production volume in modern manufacturing practices.

5.3.2.2 The Extrusion Process

The extrusion mechanism plays a pivotal role in both ME-3DP and injection molding processes, albeit with distinct applications and characteristics. In ME-3DP, the extrusion mechanism involves the controlled deposition of material in a layer-by-layer fashion to build up a three-dimensional object. This process typically utilizes a filament or resin material that is fed into an extrusion system. Within the extruder, the material is melted and forced through a nozzle with a specific diameter, determining the width of the extruded material. The nozzle moves according to the design specifications, tracing the desired paths and shapes, while the material solidifies rapidly upon extrusion, forming a solid layer. This repeated layering process results in the gradual creation of the final object. The extrusion mechanism in ME-3DP is renowned for its precision and flexibility, enabling the creation of intricate geometries and customized structures with minimal waste.

In injection molding, the extrusion mechanism takes a somewhat different form. It involves the injection of molten material into a mold cavity at high pressure and temperature. The mold comprises two halves, with one acting as a stationary base and the other as a moving component that closes over the mold cavity. Molten material is pumped into the mold cavity through a runner system and gate, filling the space that shapes the object. As the material cools and solidifies within the mold, the moving half is opened to reveal the formed part. Injection molding's extrusion mechanism ensures high repeatability and precision, producing parts with consistent dimensions, surface finish, and material properties.

While both processes utilize extrusion principles to manipulate materials, their execution and outcomes differ significantly. ME-3DP's layer-by-layer approach grants unparalleled design flexibility and the ability to create complex structures but can result in anisotropic properties due to the discrete layering. Injection molding, driven by high pressure and mold precision, is adept at producing uniform, high-quality parts at a larger scale but requires the production of molds, adding an upfront cost and time consideration.

5.3.2.3 Anisotropy and the Poor Strength in Z-direction of 3D-printed Parts

Due to the layer-wise deposition process, components produced through ME-3DP often exhibit anisotropic behavior, which correlates with the direction of printing. In general, the print platform where filaments are deposited is defined as the x–y plane, while the stacking direction of layers is defined as the Z-direction (Figure 5.2).

According to relevant research findings, the mechanical strength within the x–y plane is influenced by the raster angle. As the raster angle increases from 0° to 45°/−45° and then to 90°, the tensile strength of neat PLA components decreases from 64.3 to 54.2 MPa and eventually to 25.7 MPa [7]. This trend holds true for nearly all FFF-printed specimens, regardless of the polymer material used. In simpler terms, the tensile strength of the 90° configuration is generally lower than that of the 0° configuration within the x–y plane. The anisotropic nature of ME-3D printed parts is further evident in their mechanical properties' strong reliance on build orientation. Specifically, the strength along the z direction tends to be the weakest. As per the reported data, the tensile strength of a PLA component in the ZXY orientation measured 25.1 MPa, even slightly inferior to parts printed with a 90° raster angle within the x–y plane. This strength value represents only 47% of the maximum strength achievable for the printed part. This pattern is consistent for other mechanical properties, including flexural and impact attributes, as well as both static and dynamic mechanical characteristics.

In comparison to other components produced through AM, ME-3D printed parts exhibit the most pronounced anisotropy. The considerable

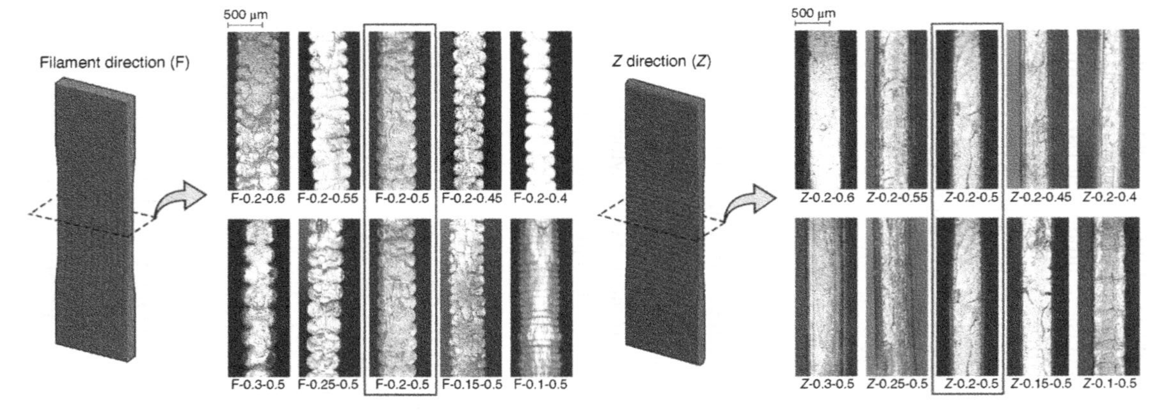

Figure 5.2 An example regarding the part distortion in 3D printing of objects. Source: Allum et al. [6]/with permission from Elsevier.

anisotropic nature of the mechanical properties within these parts undoubtedly constrains their viability for practical applications [8]. Consequently, the majority of ME-3D printed parts find utility primarily as prototypes rather than load-bearing products, primarily due to their inadequate strength and limited functionality. Addressing the deficient mechanical performance in the *Z*-direction becomes a crucial imperative to stimulate the continued advancement of ME-3DP technology. There is a common consensus that the primary factor contributing to the anisotropy of printed components arises from the arrangement of continuous material beads placed adjacently within the *x*–*y* plane. However, due to the layer-by-layer construction methodology, the deposition of unbroken material across successive layers is not feasible, thus amplifying the mechanical anisotropy along the *Z*-direction. Whenever forces are transmitted between adjacent layers in the *Z*-direction, these actions happen at distinct interfaces between materials that lie within the *x*–*y* planes [9]. Consequently, the strength in the *Z*-direction is dictated by the formation of bonding between these layers.

According to the established principles of polymer–polymer interface healing, the strength of bonding between layers is established through the diffusion of molecular chains, which is notably affected by the thermal conditions existing between these layers [10]. Considering the thermal profile of the ME-3DP process, non-isothermal interfacial healing takes place. Consequently, the initial processing conditions exert a significant influence on the bonding state of the printed parts [11]. Greater durations spent above the glass transition temperature (T_g) result in enhanced interlayer bonding strength. Simultaneously, the properties of molecular chains, particularly their relaxation behavior, also play a role in determining interlayer bonding strength [12]. In the case of semicrystalline polymers, the molecular diffusion at the interface is affected by the process of crystallization. Consequently, the interlayer bonding strength of semicrystalline polymers exhibits heightened complexity. As per certain research findings, the utilization of a slow-crystallizing PAEK polymer, as opposed to the fast-crystallizing grades, yields an improvement in z-directional strength [13].

Furthermore, the anisotropy of components is influenced by additional factors such as disentanglement effects and the alignment of fibers in fiber-reinforced composites. As highlighted in certain studies [14, 15], the strength of welds is not solely determined by the inter-diffusion of molecular chains; the arrangement of the intertwined polymer network also emerges due to the flow during printing. As a result, the overall print structure for each filament is uniform. Moreover, the fibers utilized for reinforcing the polymer matrix are oriented to align with the printing flow direction [16, 17]. Consequently, incorporating fibers enhances the mechanical properties along the 1-axis. However, the mechanical attributes along the 2-axis, which includes the *Z*-direction, either remain largely unaffected or experience a reduction due to the incorporation of reinforcing fibers.

5.3.2.4 The Lower Building Rate of ME-3DP

The issue of low building rates in the 3D printing process has emerged as a significant challenge that impacts the efficiency and feasibility of AM. The slow pace of building, where material is incrementally deposited to form each layer, contributes to prolonged production timelines, limiting the technology's potential for rapid and on-demand manufacturing. This challenge is particularly evident when creating larger or intricate parts, as the cumulative time required for layer-by-layer deposition can extend to hours or even days. Such extended production durations hinder the competitiveness of 3D printing in comparison to traditional manufacturing methods, like injection molding, which can achieve significantly faster cycle times. The reasons behind the low building rates are multifaceted. Firstly, the layer-by-layer approach inherently entails numerous stops and starts as the print head or nozzle traverses across the build area to form each layer, resulting in inherent time inefficiencies. Additionally, the materials used in ME-3DP often require precise cooling and solidification before proceeding to the next layer, contributing to the overall delay. The need for support structures, to prevent deformation during printing further, adds to the time required, as these structures need to be carefully designed and subsequently removed post-printing.

Efforts to address the low building rate challenge include optimizing printing parameters, enhancing material formulations, and exploring novel printing techniques like continuous liquid interface production (CLIP) and high-speed sintering. However, while advancements have been made, the issue of low building rates continues to be a focal point of research and development in the AM field, as increasing the speed of production without compromising quality remains a critical goal for the widespread adoption of 3D printing technology.

5.4 Future Perspective: Potential Roadmaps Toward Solving the Key Challenges of ME-3DP

The journey toward addressing the pivotal challenges inherent to the ME-3DP process has spurred the formulation of potential roadmaps, each charting a course to unlock the technology's full potential. To overcome the barriers of low building rates and anisotropic properties, one strategic path involves delving into advanced materials engineering. The exploration of novel materials with enhanced printability and optimized thermal properties holds the promise of accelerating deposition rates while mitigating anisotropy. Simultaneously, the refinement of extrusion mechanisms, leveraging multi-nozzle systems or continuous extrusion approaches, could revolutionize the pace of material deposition and lead to the creation of more isotropic parts. In tandem with material advancements, the trajectory to tackle mechanical property inconsistencies lies in the development of robust interlayer bonding techniques. Research endeavors are focused on understanding molecular diffusion at interfaces, optimizing thermal conditions, and designing advanced extruders capable of promoting stronger adhesion between layers. These efforts, combined with predictive modeling and simulation, aim to foster a deeper comprehension of the interplay between process parameters and mechanical properties, ultimately leading to more reliable and predictable outcomes.

To surmount the challenge of unpredictability in part properties, a roadmap steers toward data-driven innovation. Harnessing the potential of machine learning and artificial intelligence, researchers are working to establish correlations between process variables, material characteristics, and final part performance. By assimilating vast datasets, these methodologies promise to uncover patterns and insights, enabling the creation of robust models that can forecast part properties based on specific printing parameters.

As the ME-3DP landscape evolves, another roadmap envisions the convergence of hardware, software, and process control. Advances in real-time monitoring systems, adaptive control algorithms, and closed-loop feedback mechanisms are poised to enhance printing precision and consistency. These developments could effectively counteract deviations caused by factors such as temperature variations or material inconsistencies, promoting the creation of parts with superior accuracy and reliability. In essence, these potential roadmaps represent concerted efforts to navigate the intricate terrain of ME-3DP challenges. Through the synergy of innovative materials, advanced interlayer bonding techniques, data-driven insights, and refined process control, the path forward envisions a future where ME-3DP not only overcomes its obstacles but also transforms industries with unprecedented speed, precision, and versatility. Accordingly, some important features of this objective are presented in this section.

5.5 High Building Rate ME-3DP Process

The building rate in ME-3DP, a crucial factor determining the speed of the AM process, is inherently constrained by the capabilities of the printing system's module. In this context, enhancing the performance of individual modules within the system holds the potential to yield substantial improvements in the overall building rate of ME-3DP [9, 18]. This acknowledgment underscores the significance of focusing on module-level advancements to expedite the AM process.

Presently, two promising avenues emerge as potential technologies capable of realizing higher building rates within the ME-3DP process. These avenues represent distinct approaches that could potentially revolutionize the speed and efficiency of the technology. By addressing the challenges inherent to low building rates, these technologies aim to enhance the competitiveness of ME-3DP as a viable manufacturing method across various industries. As the pursuit of faster AM continues, these innovative pathways promise to reshape the landscape of ME-3DP, potentially catapulting the technology into new realms of application and significance.

5.6 Big Area Additive Manufacturing (BAAM) System

BAAM stands out as a notable advancement in the realm of large-format printing systems, pioneered by Oak Ridge National Laboratory. Designed to produce structures on a scale of several meters, this technology introduces the capability to swiftly print sizable components while maintaining high extrusion rates [19]. The unique approach of utilizing feedstock in the form of thermoplastic or reinforced thermoplastic pellets brings about a substantial reduction in material deposition costs, presenting a cost-efficient alternative that slashes expenses by a factor of 20 times. At the core of this system lies a single-screw extruder, responsible for both melting the plastic pellets and subsequently extruding the molten material onto a heated print bed. This single-screw extruder further distinguishes itself by its remarkable deposition speed, surpassing conventional polymer AM systems by over 200 times and boasting rates of up to 50 kg/h.

An additional feature contributing to BAAM's prowess is its expansive build platform, capable of accommodating structures of impressive dimensions. With the potential to construct objects as substantial as 6 meters in length, 2.4 m in width, and 1.8 m in height, BAAM's build capacity surpasses most commercial systems by a factor of approximately

10. This expanded envelope of production capability positions BAAM as a technology with promising implications across diverse sectors, including automotive, aerospace, and energy industries. The exceptional attributes of BAAM, from its rapid extrusion rates and economical material usage to its capacity for fabricating sizable structures, underscore its potential to revolutionize manufacturing methodologies, making substantial contributions to the realms of transportation, aeronautics, and power generation.

5.7 Faster FFF 3D Printing System

To surmount the constraints imposed by performance-limiting mechanisms, the development of a swift FFF system hinges on several key enhancements. These enhancements encompass an improved material feed mechanism, augmented heat transfer rates from the liquefier wall to the filament core, and the optimization of gantry positioning systems guided by open-loop stepper motors. Researchers' estimations suggest that a remarkable 10-fold increase in building rate compared to typical commercial desktop systems could be attainable through the enhancement of printer module performance and the implementation of refined toolpath planning algorithms [19]. In parallel, innovative hardware components have emerged to bolster the potential of swift FFF systems. Extruders featuring multipoint contact with the filament have been introduced to the market, ensuring a firm grip on the filament and preventing slippage during the printing process. Furthermore, the liquefier, now equipped with dual heat zones and meticulous temperature control mechanisms, maintains the polymer at a consistent temperature throughout the extrusion process. The incorporation of step-motor driving systems aids in achieving elevated velocity and acceleration for the stepper motor, ultimately allowing for printing speeds reaching up to 250 mm/s.

Recently, a rapid FFF system boasting a building rate of approximately 720 cm^3/h has become available on the market. Looking ahead, it is foreseeable that further advancements in module functionality will pave the way for even swifter FFF systems. The integration of AI-driven assistance systems, such as camera recognition, in conjunction with refined path planning algorithms, holds promise for significantly elevating building rates. As these technological advancements continue to converge, it is reasonable to anticipate the emergence of fast FFF systems with progressively higher building rates, propelling the field of AM into a new era of enhanced speed and efficiency.

5.8 Improvement of Interfacial Bonding and Strength in *Z*-direction

Examining the underlying factors influencing the weakened bond interface in the context of ME-3DP underscores the need for a dual-pronged approach encompassing both process control and judicious material selection. To effectively bolster the interfacial bonding strength and concurrently diminish the pronounced anisotropy characteristic of printed components, these two key aspects must be thoughtfully addressed. Process control plays a pivotal role in sculpting the final quality of ME-3D printed parts. Fine-tuning various printing parameters, such as extrusion temperature, layer height, and print speed, can exert a significant influence on the formation of robust interlayer bonds. By meticulously optimizing these parameters, the thermal conditions during printing can be adjusted to facilitate enhanced molecular chain diffusion at the interface, leading to stronger and more dependable bonds between successive layers. Process control also extends to real-time monitoring and feedback mechanisms, allowing for dynamic adjustments during printing to counteract potential deviations and ensure consistent interfacial bonding throughout the entire production process.

Simultaneously, the choice of material assumes paramount importance in augmenting interfacial bonding strength. The selection of polymers with compatible melt viscosities and thermal properties can foster more cohesive bonding between layers. Additionally, the incorporation of additives, such as coupling agents or compatibilizers, can promote molecular adhesion and alignment, further enhancing the overall interlayer adhesion. Furthermore, materials with balanced mechanical properties and thermal behaviors across multiple directions can contribute to reducing anisotropy, resulting in more uniform and predictable mechanical performance across all orientations.

Harmonizing process control and material selection leverage a comprehensive strategy to surmount the challenge of weak bond interfaces and the ensuing anisotropy in ME-3D printed parts. This combined effort, founded on precise control over the printing process parameters and the strategic utilization of materials with optimal properties, offers a pathway toward producing components with heightened strength, reliability, and uniformity across diverse orientations.

The inadequate adhesion between layers stems from the limited diffusion of large and sluggish polymer chains across the interface between filaments. Enhancing diffusion and interlayer adhesion in FDM-fabricated parts can be facilitated by employing filament materials that contain lower molecular weight components. This choice is driven by the fact that polymers with lower molecular weights exhibit quicker diffusion across the inter-filament interface, thereby fostering improved interlayer adhesion [20]. A promising solution to address the weakness in the z-direction involves utilizing filaments with a core-shell structure, a concept realized through polymer co-extrusion [21–24]. Typically, the filament's core possesses a notably higher glass transition temperature than its sheath, imparting favorable attributes such as enhanced printability, stiffness, and dimensional stability. The interplay between these dual materials benefits from well established chemical compatibility, often resulting in a phase-separated blend characterized by excellent adhesion between phases. Predominantly, the most prevalent combination involves a dual-material filament comprising ABS with a PC core. This combination is frequently subjected to annealing post-printing, a treatment that reinforces bonding between printed layers while upholding dimensional stability.

Compared to conventionally printed single-material filaments, components produced using the dual-material filament exhibit remarkable enhancements. For instance, post-printing annealing culminates in a 5-fold amplification in z-direction impact toughness and a 4-fold increase in z-direction tensile strength. Impressively, these benefits are achieved without compromising part dimensional accuracy or surface quality. Notably, even when elevated printer nozzle temperatures are employed, the inherent stability of the core material exerts a stabilizing effect, ensuring excellent dimensional precision and surface finish.

5.9 Conclusions

3D printing process holds immense promise for revolutionizing manufacturing and design paradigms across industries. However, a range of challenges currently impedes its seamless integration and optimal utilization. The intricate interplay between material properties, printing technologies, and design complexity underscores the need for holistic solutions. Addressing material compatibility, refining printing parameters, and enhancing post-processing techniques are pivotal steps to overcome the hurdles of accuracy, speed, and part integrity. Moreover, as 3D printing evolves, researchers, engineers, and industry practitioners must collaboratively innovate to unlock the technology's full potential. By tackling these challenges head-on, the 3D printing landscape can mature into a transformative force, enabling customized, efficient, and sustainable manufacturing processes that redefine how we create objects and push the boundaries of what is achievable.

References

1 Crump, S.S., Apparatus and method for creating three-dimensional objects. 1992, Google Patents.

2 Hiemenz, J. (2011). *3D printing with FDM: How it Works*, vol. 1, 1–5. Stratasys Inc.

3 Verhoef, L.A. et al. (2018). The effect of additive manufacturing on global energy demand: an assessment using a bottom-up approach. *Energy Policy* 112: 349–360.

4 Vanaei, H.R., Khelladi, S., and Tcharkhtchi, A. (2022). Roadmap: numerical-experimental investigation and optimization of 3D-printed parts using response surface methodology. *Materials* 15 (20): 7193.

5 Lalegani Dezaki, M., Mohd Ariffin, M.K.A., and Hatami, S. (2021). An overview of fused deposition modelling (FDM): research, development and process optimisation. *Rapid Prototyping Journal* 27 (3): 562–582.

6 Allum, J. et al. (2020). Interlayer bonding has bulk-material strength in extrusion additive manufacturing: new understanding of anisotropy. *Additive Manufacturing* 34: 101297.

7 Gao, X. et al. (2021). Fused filament fabrication of polymer materials: a review of interlayer bond. *Additive Manufacturing* 37: 101658.

8 Guessasma, S. et al. (2016). Anisotropic damage inferred to 3D printed polymers using fused deposition modelling and subject to severe compression. *European Polymer Journal* 85: 324–340.

9 Go, J. et al. (2017). Rate limits of additive manufacturing by fused filament fabrication and guidelines for high-throughput system design. *Additive Manufacturing* 16: 1–11.

10 Coasey, K. et al. (2020). Nonisothermal welding in fused filament fabrication. *Additive Manufacturing* 33: 101140.

11 Sun, Q. et al. (2008). Effect of processing conditions on the bonding quality of FDM polymer filaments. *Rapid Prototyping Journal* 14 (2): 72–80.

12 Gilmer, E.L. et al. (2021). Temperature, diffusion, and stress modeling in filament extrusion additive manufacturing of polyetherimide: an examination of the influence of processing parameters and importance of modeling assumptions. *Additive Manufacturing* 48: 102412.

13 Yi, N. et al. (2021). Slow and fast crystallising poly aryl ether ketones (PAEKs) in 3D printing: crystallisation kinetics, morphology, and mechanical properties. *Additive Manufacturing* 39: 101843.

14 Costanzo, A. et al. (2020). Residual alignment and its effect on weld strength in material-extrusion 3D-printing of polylactic acid. *Additive Manufacturing* 36: 101415.

15 McIlroy, C. and Olmsted, P. (2017). Disentanglement effects on welding behaviour of polymer melts during the fused-filament-fabrication method for additive manufacturing. *Polymer* 123: 376–391.

16 Liao, G. et al. (2018). Properties of oriented carbon fiber/polyamide 12 composite parts fabricated by fused deposition modeling. *Materials & Design* 139: 283–292.

17 Jiang, D. and Smith, D.E. (2017). Anisotropic mechanical properties of oriented carbon fiber filled polymer composites produced with fused filament fabrication. *Additive Manufacturing* 18: 84–94.

18 Go, J. and Hart, A.J. (2017). Fast desktop-scale extrusion additive manufacturing. *Additive Manufacturing* 18: 276–284.

19 Duty, C.E. et al. (2017). Structure and mechanical behavior of Big Area Additive Manufacturing (BAAM) materials. *Rapid Prototyping Journal* 23 (1): 181–189.

20 Levenhagen, N.P. and Dadmun, M.D. (2017). Bimodal molecular weight samples improve the isotropy of 3D printed polymeric samples. *Polymer* 122: 232–241.

21 Ouassil, S.E. et al. (2023). Investigating the effect of printing conditions and annealing on the porosity and tensile behavior of 3D-printed polyetherimide material in Z-direction. *Journal of Applied Polymer Science* 140 (4): e53353.

22 Peng, F. et al. (2019). 3D printing with core–shell filaments containing high or low density polyethylene shells. *ACS Applied Polymer Materials* 1 (2): 275–285.

23 Hart, K.R., Dunn, R.M., and Wetzel, E.D. (2020). Tough, additively manufactured structures fabricated with dual-thermoplastic filaments. *Advanced Engineering Materials* 22 (4): 1901184.

24 Koker, B. et al. (2022). Enhanced interlayer strength and thermal stability via dual material filament for material extrusion additive manufacturing. *Additive Manufacturing* 55: 102807.

6

Importance of Multi-objective Evaluation in 3D Printing

Kasin Ransikarbum[1,2] *and Namhun Kim*[3]

[1] *Department of Industrial Engineering, Ubonratchathani University, Ubonratchathani, Thailand*
[2] *Department of Industrial Engineering, Kasetsart University, Bangkok, Thailand*
[3] *Department of Mechanical Engineering, Ulsan National Institute of Science and Technology, Ulsan, Republic of Korea*

6.1 Introduction

Additive manufacturing (AM) or three-dimensional printing (3DP) is used interchangeably, which refers to the method of engineering and manufacturing objects layer-by-layer, part-by-part from 3D digital files, using computer-aided design software. In contrast to traditional/subtractive manufacturing techniques, 3DP has offered benefits toward greater design flexibility, lesser material, lead time improvement, and green manufacturing. Thus, manufacturers in the industry acknowledged 3DP as a capable method and began employing 3D printers in not only their research and development (R&D) and prototype design but also their manufacturing processes for end-use tools and parts. Other jargon commonly used for AM also includes direct digital manufacturing (DDM), rapid tooling (RT), rapid manufacturing (RM), and solid free form (SFF) fabrication [1].

The American Society for Testing and Materials [2], in particular, has classified AM based on relevant technologies into seven groups: (i) photopolymer vat, (ii) material extrusion, (iii) powder bed fusion, (iv) directed energy deposition, (v) sheet lamination, (vi) material jetting, and (vii) binder jetting. The survey from Wohlers [1] suggested that fused deposition modeling (FDM) – material extrusion category, stereolithography (SLA) – photopolymer vat category, and selective laser sintering (SLS) – powder bed fusion category, are among the top three AM technologies based on the number of installed industrial systems worldwide. To date, various researchers have addressed challenges facing AM quality,

Industrial Strategies and Solutions for 3D Printing: Applications and Optimization, First Edition. Edited by Hamid Reza Vanaei, Sofiane Khelladi, and Abbas Tcharkhtchi.

production and process planning, and supply chain evaluation in the literature [3]. A part of the challenges of AM is owing to conflicting behavior of criteria/objectives relevant to 3D printed parts' properties as well as 3DP's process and production planning.

Production and process problems for 3DP can be regarded as a type of multi- objective/criteria problem, given conflicting behavior among objectives/criteria of 3D printed parts as well as planning requirements. We note that the term objectives and criteria are similarly used in the context of production planning in the literature. For instance, improving desired property and quality of a 3D printed part (more is better) typically requires higher production cost and time (less is better), which is hence conflicting. Thus, multi-objective/criteria evaluation becomes a vital decision-making tool through a capability to deal with practical and real-world 3DP production and process problems.

In contrast to a single-objective/criterion problem, in which the superiority of a solution can be easily determined, the multi-objective/criteria problem involves two or more objectives/criteria that are typically conflicting and need to be considered simultaneously. In fact, multi-objective programming (MOP) can be viewed as a subarea of multi-criteria decision analysis (MCDA), which encompasses mathematical problems involving more than one objective function to be assessed. Besides, a number of well known MCDA tools to assess criteria weight and/or to select the best alternative among conflicting criteria are, for example, analytic hierarchy process (AHP), data envelopment analysis (DEA), technique for order of preference by similarity to ideal solution (TOPSIS), and fuzzy logic, normalization. Some of these tools are to be selectively exemplified in this chapter.

In this chapter, we discuss the multi-objective/criteria evaluation for 3DP, which can analyze trade-offs among key objectives/criteria relevant to various aspects of 3D printed parts and processes. In addition, most AM process activities are found to be planned by human planners, and AM production activities are mainly designed based on the intuition and experience of operators. Hence, a lack of systematic planning can cause under-utilization, thus further damaging the impact of implementing 3DP within an organization. Therefore, we illustrate the current state of multi-objective evaluation of 3DP using a framework of DSS for 3DP inclusive of part orientation, printer selection, and part-to-printer assignment modules in our chapter. Several applications and illustrations in each module of the DSS present how trade-offs exist among multiple criteria/objectives of the problems and how they can be evaluated in the context of decision analysis.

The remaining sections of this chapter are organized as follows. We overview the current state of multi-objective/criteria evaluation of 3DP in Section 6.2. Then, the illustrated DSS for 3DP under multiple-objective/criteria evaluation is presented in Section 6.3. Section 6.4 next presents discussion and managerial insights. Finally, Section 6.5 outlines our chapter conclusions and recommendations.

6.2 The Current State of Multi-Objective Evaluation of 3DP

Challenges of the multi-objective/criteria modeling approach can be evidently traced to the traditional supply chain as evidenced by a large number of studies in the literature. The flow of traditional supply chain starts from raw materials from suppliers at the upstream, work-in-process (WIP) to manufacturing facilities at the midstream, finished goods (FG) to distribution centers, and sales to retailers for consumers at the downstream. Some future research recommendations concerning multi-objective/criteria evaluation for traditional supply chains are, for example, multi-objective assessment of the sustainability paradigm, solution approach for multi-objective meta/heuristics, and risk consideration as a part of the multi-objective model.

In contrast to traditional supply chains, AM or 3DP supply chain evaluation is more challenging and complex. The first tier of the supply chain for 3DP involves also web-based retailers, where an end customer orders a 3D-designed part via an online transaction. Thus, the supply chain structure shifts from the traditional network toward a retail–production–distribution model. Production scheduling and logistical planning become more perplexing for a complex 3DP supply chain with various types of part requirements, material types, and diverse 3DP technologies. Trade-offs among key objectives/criteria relevant to various characteristics of 3D printed parts and process planning also suggest that multi-objective tools relevant to MCDA are also needed for proper analysis.

The strategic, tactical, and operation planning levels of an AM/3DP supply chain is shown in Figure 6.1. That is, the 3DP supply chain's strategic level planning at a macro level and the proposed tactical/operation level at the production planning stage is illustrated. In particular, the production planning stage consists of three related modules, which are part orientation, printer selection, and part-to-printer assignment. That is, the proper orientation of each part needs to be decided. Next, printer alternatives should be examined and designated for the 3DP task. Then,

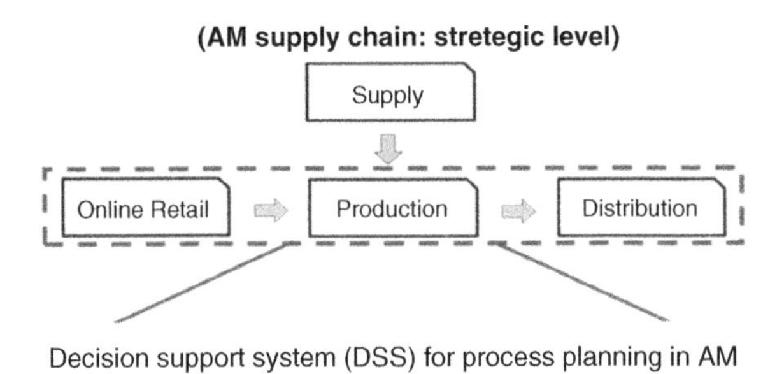

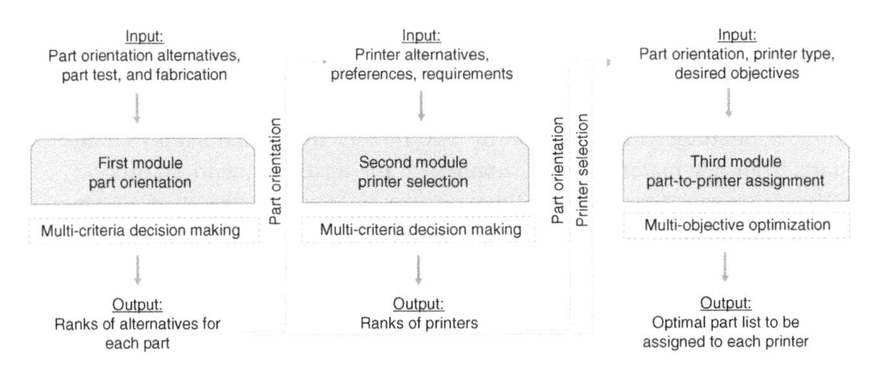

Figure 6.1 A framework of decision support system for 3DP production planning.

oriented parts are assigned to a batch of 3D printers based on printers' properties, part characteristics, customer requirements, and so on. In this chapter, our focus is to present how the MCDA techniques inclusive of MOP, under conflicting objectives/criteria, are applied in aiding production planning stage for 3DP.

6.2.1 Part Orientation Problem in 3DP

The decision for an orientation selection is an integral factor of production and process planning in 3DP. This process is typically followed by a number of critical steps, including support generation, toolpath designation, layer manufacturing, and post-processing. A number of studies show diverse methodologies to tackle the orientation problem for varied 3DP types. These researchers' studies similarly point out that the part orientation decision is not a trivial task as satisfying one criterion of the problem may adversely affect some other criteria of interest. Some recent and prominent studies highlighting existing orientation problems are briefly reviewed next.

Early studies relevant to the 3DP part orientation problem typically use a single and simple method to investigate a single or a few criteria of orientation problem in 3DP without investigating the complex problem with varied MCDM methods. Over the last five years, research directions for 3DP orientation problem have integrated various MCDA techniques and optimization procedures into their study. Both quantitative and qualitative criteria are also explored to take decision-makers' perspectives into account under multiple criteria. For example, Zhang et al. [4] developed non-supervised machine learning to analyze the orientation problem of complex models for a healthcare application. The authors evaluate surface roughness, support volume, and facet clusters for electron beam melting (EBM) in their study. Di Angelo et al. [5] proposed the multi-objective optimization model to investigate the orientation problem for FDM using varied case studies. In addition, Leirmo and Martinsen [6] proposed the feature recognition method to assess the impact of the staircase effect on the orientation problem from varied parts for SLS. Additionally, Ransikarbum et al. [7] developed the integrative framework of MCDA for part orientation analysis in FDM and SLS case study with varied conditions/criteria. Lately, Yang et al. [8] has proposed the integrated double-layer priority aggregation and MCDA tool to assess part orientation under surface precision with feature, printing time, layout area, and support volume criteria.

6.2.2 Printer Selection Problem in 3DP

The 3D printer selection is also considered one of the critical decisions that can affect performance use and business perspective. This problem is compounded by the fact that a number of 3D printer companies and third-parity material providers have increased with diverse 3DP technologies, varied material suppliers, and various printer sizes [1, 7]. The following selective studies emphasize that 3D printer selection can be viewed as a type of MCDA problem with multiple criteria.

Recent studies have been proposed to tackle 3D printer selection decision and the trend of incorporating uncertain aspect from decision-makers have been observed. For example, Peko et al. [9] used the integrated preference ranking organization method for enrichment of evaluations (PROMETHEE) technique and fuzzy AHP (FAHP) to evaluate FDM, SLS, and PolyJet printers. Criteria concerning cost, time, accuracy, surface roughness, mechanical property, and post-processing are evaluated using the fuzzy logic in their work. In addition, Prabhu and Llangkumaran [10] proposed the hybrid MCDA technique, in which the FAHP, elimination and choice expressing reality (ELECTRE), and Viekriterijumsko Kompromisno

Rangiranje (VIKOR), are used to evalusate desktop AM printer selection for automotive parts. Relevant criteria with respect to printer cost, filament cost, volume, printing speed, layer thickness, and extruder type are discussed. Chen and Wu [11] applied the FAHP technique alone to evaluate the 3D printer selection problem. The authors use the center-of-gravity (COG) technique to obtain aggregated group decision-making in their work. Moreover, Ransikarbum and Khamhong evaluated the 3D printer selection problem for healthcare applications using integrative MCDA techniques inclusive of FAHP and TOPSIS in their study. Lately, Raja and Rajan [12] have evaluated the FDM printer selection using the Fuzzy TOPSIS technique in their analysis.

6.2.3 Part-to-Printer Assignment Problem in 3DP

In evaluating the feasibility of the part-to-printer assignment/scheduling process, an important requirement is the ability to compose batches of parts accounting for various objectives/criteria, such as available space in the 3D printers, operation cost and time, customer requirements, and load balances among printers. Recent studies have proposed varied optimization and meta/heuristic approaches for improving 3D printer utilization and part-to-printer assignment problems by taking part orientation into account. Some prominent studies concerning part-to-printer assignment problem are reviewed next.

According to Manogharan et al. [13], the impact of AM production cost and time can be mitigated by increasing the batch size. That is, by nesting multiple unique parts, the unit cost of each part is expected to be lowered. However, existing studies with one printer or one criterion consideration lack a perspective of shared 3D printers with multiple printers and diverse criteria in the real business platform as well as in the future-growth condition. In fact, existing studies largely consider only a single economic objective without exploring trade-offs among other objectives. Moreover, accounting for various 3DP technologies in the model can reveal some insights for interested decision-makers. Recently, Ransikarbum et al. [14] proposed a decision-support tool that integrates production and distribution planning in AM involving FDM, SLA, and SLS. An integrated MCDA tool is proposed, such that the multi-objective optimization was initially developed to schedule component batches to a network of AM printers. Then, the AHP technique was applied to analyze trade-offs among conflicting criteria/objectives in their study. Moreover, Altekin and Bukchin [15] proposed the biobjective optimization for direct metal laser sintering. Both cost and makespan minimization are assessed in their work. In addition,

Tafakkori et al. [16] proposed the metaheuristic optimization with an aim to take into account three aspects of the sustainability paradigm for 3DP planning into account in their model.

We next summarize key discussed studies involving part orientation, printer selection, part-to-printer assignment, and key production planning problems for 3DP's DSS as shown in Table 6.1. We note that the listed studies are not meant to be inclusive, but provide an illustrative trend of multi-objective evaluation studies in this 3DP research area.

6.3 Decision Support System for 3DP Under Multi-Objective Evaluation

A decision support tool can be used to aid a decision maker/planner in efficiently and effectively planning for production activities relevant to 3DP. We next illustrate the DSS framework for 3DP under multi-objective/criteria evaluation. Initially, the 3D part orientation problem is demonstrated, in which a number of deterministic MCDA tools can be applied to examine the best orientation among conflicting criteria of interest. Next, we assess the 3D printer selection problem under various criteria using MCDA approach. The stochastic MCDA approach is presented to demonstrate how uncertainty can be taken into account under an evaluation of multiple criteria in making decision. Finally, the integrated MOP model and AHP technique are exemplified to present trade-off analysis among multiple objectives for the part-to-printer assignment/scheduling problem.

6.3.1 Part Orientation

We initially illustrate the part orientation case study in this subsection following the key publication of Ransikarbum et al. [7] to demonstrate how multiple criteria can be incorporated and assessed. That is, the integrative framework of MCDA for part orientation analysis in 3DP is investigated, in which both quantitative and qualitative type criteria are considered. The quantitative data are initially examined using the DEA technique without preferences from a decision-maker. Then, a decision-maker's preferences are qualitatively analyzed using the AHP technique. Next, the proposed framework combining explicit data as in DEA, implicit preference as in AHP, and linear normalization (LN) technique is applied, which can support decision-making for 3DP part orientation. Two particular AM technologies, FDM and SLS, are demonstrated as a case study.

Table 6.1 Summary of current research for multiple-criteria/objective evaluation of 3DP.

Study	Method/technique	AM technology	(Number) criteria/ objectives
	(Module 1) part orientation		
Zhang et al. [4]	Non-supervised machine learning	EBM	(3) Surface roughness, support volume, facet clusters
Qin et al. [17]	Fuzzy MCDM	SLA, SLS	(5) Surface roughness, time, cost, support volume, favorableness
Di Angelo et al. [5]	Multi-objective optimization	FDM	(2) Cost, surface quality
Leirmo and Martinsen [6]	Feature recognition	SLS	(1) Staircase effect
Ransikarbum et al. [7]	Integrative MCDM (DEA, AHP, LN)	FDM, SLS	(6) Part cost, time, surface quality, part accuracy, support volume, mechanical property
Yang et al. [8]	Double-layer priority aggregation MCDM	General AM	(4) Surface precision with feature, printing time, layout area, support volume
	(Module 2) printer selection		
Peko et al. [9]	FAHP, PROMETHEE	FDM, SLS, PolyJet	(6) Accuracy, surface roughness, mechanical property, cost, time, post-processing
Prabhu and Llangkumaran (2019)	FAHP, GRA-TOPSIS	Desktop AM	(6) Build volume, speed, layer thickness, extruder, machine cost, and support material cost.

Table 6.1 (Continued)

Study	**Method/technique**	**AM technology**	**(Number) criteria/ objectives**
Netto et al. [18]	AHP	Desktop AM	(7) Technical (surface roughness, dimensional accuracy, geometric accuracy, build time), software (estimated time, estimated material), cost
Chen and Wu [11]	FAHP	Desktop AM	(6) Price, rating, application types, build area, resolution, material types
Ransikarbum and Khamhong [19]	FAHP, TOPSIS	FDM	(11) Product (accuracy, surface finish, build cost, build time, appearance), material (cost, tensile strength, surface finish), printer (cost, build capacity, user preference)
Raja and Rajan [12]	Fuzzy TOPSIS	FDM	(9) Price, build volume, extruder type, printing speed, operating temperature, filament material, tolerance, environmental factor, safety
	(Module 3) part-to-printer assignment		
Ransikarbum et al. [14]	Multi-Objective Optimization	FDM, SLA, SLS	(4) Total operating cost, load balance, total tardiness, total number of unprinted parts
Altekin and Bukchin [15]	Biobjective Optimization	Direct metal laser sintering	(2) Cost, makespan
Tafakkori et al. [16]	Meta-heuristic algorithm	General AM	(3) Profit, energy utilization of machines, goodwill losses

In order to demonstrate how MCDA tools can be used to recommend solutions for multiple-criteria 3DP's orientation problem, selective MCDA tools are further discussed as follows.

6.3.1.1 Data Envelopment Analysis (DEA)

DEA is a type of multi-criteria productivity analysis model that compares each variable of interest with the best-performing one. Variables in DEA analysis are often referred to as decision-making units (DMUs), in which the main aim is to provide benchmarking guidelines for inefficient DMUs. Advantages of DEA include the capability to handle multiple input and output criteria, where the sources of inefficiency can be analyzed and quantified for every evaluated unit. Also, DEA allows inter-criteria comparison with varied units and a range of criteria. The relative efficiency of a particular DMU can be acquired by solving the mathematical optimization model of the DEA. Analyzed DMUs will then be considered efficient if they obtain a score of one, whereas scores that are lesser than one imply relative inefficiency. Further, more than one alternative may be found to be efficient, which can serve as benchmarking option(s) for inefficient DMUs to follow (e.g. to increase output while fixing the same input or to decrease input while obtaining the same output). The DEA technique can noticeably be used to evaluate 3DP planning under various criteria that can be classified as input or output to find the best orientation under consideration.

6.3.1.2 Analytic Hierarchy Process (AHP)

AHP is in essence based on three main sequences – hierarchy construction, priority analysis, and consistency verification. Initially, the logical hierarchy can be made such that a decision-maker or a group of decision-makers can systematically assess alternatives' priority by making pairwise comparisons among criteria and their respective alternatives concerning each criterion. AHP also allows the assessment of judgment inconsistency, which is the main distinctive contribution of the AHP when compared to other techniques. The relative importance scale (e.g. 1–9 scales) between two alternatives is dominantly used for comparison between two elements. However, other scales also exist. That is, the 1–9 fundamental scale can be translated as an equal preference or 1, moderately preferred or 3, strongly preferred or 5, very strongly preferred or 7, and extreme preference or 9, respectively with 2, 4, 6, and 8 as intermediate values. We note that the AHP technique can be used for both evaluating relative, global weight of multiple criteria as well as assessing alternatives' ranking concerning 3DP planning for the part orientation.

6.3.1.3 Linear Normalization (LN)

In multiple criteria/objective problems, all the criteria of interest may have different units with varied magnitudes. Thus, it may become a concern in decision-making as a relative rating of alternatives may change simply because of different units of measurement. Thus, a number of scalarization/normalization techniques are proposed to tackle the MCDA problem to allow inter-criterion comparison. We illustrate one well known normalization technique called LN, which can normalize different units of criteria of interest whether these criteria are benefit type (i.e. a decision-maker prefers more of it, or more is better) or cost type (i.e. a decision-maker prefers less of it, or less is better). In particular, the LN technique employs the ideal/utopia and anti-ideal/nadir solutions, where the ideal/utopia and anti-ideal/nadir solutions are the best and the worst possible alternatives for considering each criterion, respectively, in order to convert data for all alternatives concerning each criterion to be in a range between 0 and 1 and in the format of maximum desire. Thus, incorporating the LN technique for multiple-criteria problems related to 3DP can properly support inter-criteria evaluation of the part orientation.

6.3.1.4 Illustrative Case Study for Part Orientation

To illustrate how the above selective MCDA tools (i.e. DEA, AHP, and LN) can be sequentially applied for the 3DP part orientation problem under multiple criteria, the three-phase general framework is proposed and discussed as follows.

The first phase is to cautiously select alternatives and criteria for the problem of interest. The concerned problem may come from diverse decision-makers and cover different variables in terms of alternatives and criteria lists. Thus, numerous groups of data can create various kinds of rankings. We note that a ranking implies a ranked list resulting from comparing objects using specific evaluation methods with one particular criterion or with multiple criteria. From the DEA perspective, the criteria may be viewed as the input or output criteria, whereas the overall efficiency score can be treated as the outcome of the measurement. Besides, it is important to note that the DEA technique does not incorporate the experience and preferences of decision-makers for the preferred criteria in the ranking.

The second phase is to make clustering DMUs for further DEA analysis. DEA is a nonparametric method that can be used not only to measure the relative efficiency but also to designate a reference target for an inefficient DMU. Thus, DEA is a type of clustering technique that separates DMUs into two categories, efficient and inefficient. Also, as the DEA technique shows relative efficiency among peers of evaluation, data in an experiment of interest may be organized into subgroups, such that the DEA is used to analyze each subgroup rather than the entire group for practical discussion. Next, an evaluation of the criteria weight can be further assessed using AHP. Although the DEA technique can be used to assess the efficiency of a particular DMU, this method alone does not incorporate the viewpoint or experience of decision-makers for the preferred criteria in the ranking. Thus, the AHP technique can be used to secure criteria weight based on preference and experience of the decision-maker in making criteria judgments. However, unlike using judgment for criteria evaluation, using AHP to rank all alternatives can help to transform quantitative data for all alternatives of each criterion into a judgment scale during pairwise comparisons.

The final third phase is to scale measurement units using LN. This phase of normalizing actual data is important to ensure that the relative rating of alternatives will not be changed merely because of dissimilar measurement units. In particular, the LN technique can be perceived as a method to combine explicit data as in DEA, and implicit preference as in AHP to utilize both preference and objective data in supporting decision-making under various 3DP environments. Thus, by combining the criteria weight obtained from the AHP with the normalized data from LN in this phase, the list of alternatives can be ranked and interpreted. Also, the ranked list will reflect the viewpoint and judgment of a decision-maker through the criteria weight that likely differs among decision-makers involved in a decision-making process.

We next present a discussion of an experimental design of a selected case study for the part orientation problem. The test part model with a hole feature comprises of six alternatives for build directions/orientations as illustrated in Figure 6.2. In addition, six criteria are illustrated, which are build time (BT), build cost (BC), surface quality (SQ), part accuracy (PA), mechanical properties (MP), and support volume (SV) for two different AM technologies (i.e. FDM and SLS). In particular, the BT criterion refers to the time spent on layer scanning dependent on the number of slices. As orientation of the part will affect also a part's height, it follows that different orientations can greatly impact the build time. Besides, the BC criterion refers to the resources consumed during the manufacturing of a part, which usually contains direct and indirect costs. As the indirect cost can be estimated based on the build time, orientation of a part will have a substantial impact on the part cost. Next, concerning the SQ criterion, parts usually being parallel or perpendicular to the build orientation will likely have a better surface roughness than those whose face has an angle to the build direction. In contrast, declining faces resulting from an orientation will be affected by the stair-step effect when printing. Next, the PA criterion refers to the difference between

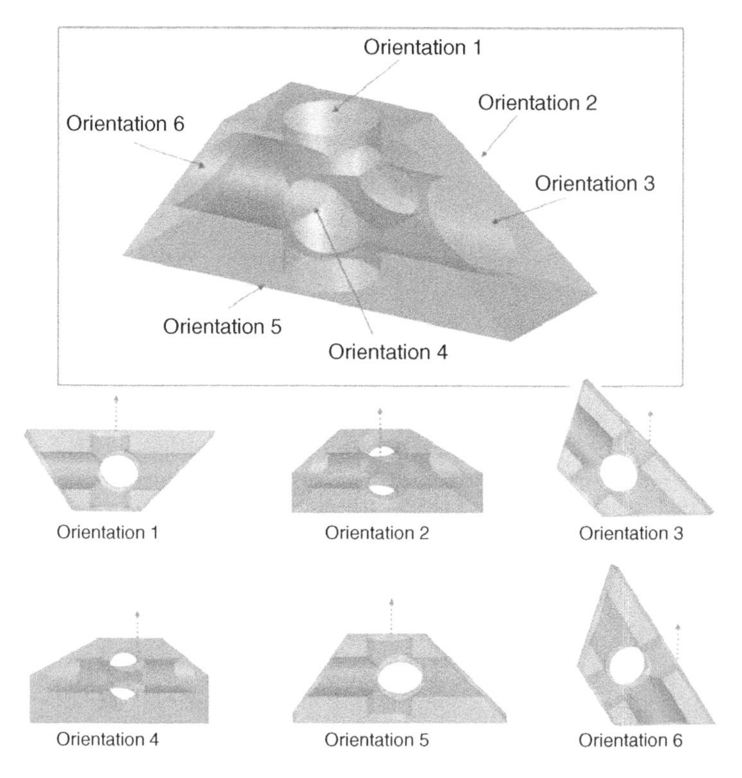

Figure 6.2 Orientation alternatives of the part orientation case study.

the produced part and the designed model. Part orientation can affect both shrinkage and distortion, which are the main factors in 3DP resulting in this inaccuracy. Moreover, it is well known that the properties of a part produced by 3DP are anisotropic. That is, orientation direction will tend to affect various mechanical properties. Last but not least, the SV criterion depends on the support structure, which is also dependent on varied 3DP technologies. For example, whereas support structure is needed in FDM for over-hanging, it is not required in SLS, as unsintered materials act as a support.

Collected data of each tested orientation under different criteria is demonstrated in Table 6.2. Data from different AM technologies and altered orientation alternatives show conflict with each other. For example, while the build time and cost for different alternatives in FDM are found to be in a similar range, SLS is shown to have varied build time and cost. This is due to the difference between the energy source and material between them. Besides, although the support material is not required in SLS, it is typically needed in FDM. The mechanical properties are also clearly rated differently between

Table 6.2 Summary of data in an experimental design.

FDM	Alt. 1	Alt. 2	Alt. 3	Alt. 4	Alt. 5	Alt. 6
BT (hr)	4.5–5.0	5.0–5.5	4.5–5.0	4.0–4.5	4.5–5.0	4.5–5.0
BC ($)	$12	$15	$13	$12	$13	$13
SQ (R_a)	R_a 5.58	R_a 5.45	R_a 13.06	R_a 2.75	R_a 9.98	R_a 11.22
PA (RMS in mm.)	0.128	0.127	0.146	0.109	0.103	0.141
MP (*good 5 bad 1*)	Score 2	Score 3	Score 1	Score 5	Score 4	Score 1
SV (g)	3	1	5	1	3	4
SLS	**Alt. 1**	**Alt. 2**	**Alt. 3**	**Alt. 4**	**Alt. 5**	**Alt. 6**
BT (hr)	4.0–4.5	5.0–5.5	6.0–6.5	3.5–4.0	4.0–4.5	7.5–8.0
BC ($)	$70–80	$75–85	$80–90	$50–60	$70–80	$90–100
SQ (R_a)	R_a 17.54	R_a 25.67	R_a 36.68	R_a 31.49	R_a 11.72	R_a 35.63
PA (RMS in mm.)	0.211	0.283	0.271	0.232	0.244	0.249
MP (*good 5 bad 1*)	Score 5	Score 3	Score 1	Score 4	Score 2	Score 1
SV (g)	No need	No need	No need	No need	No need	No need

Remark: R_a is the roughness parameter; RMS is an error between CAD file and printed part.

SLS and FDM. The total height of the oriented part will also technically affect the anisotropic property of a part. That is, there exists a higher probability of inhomogeneous density affected by gravity in a taller part. The part produced in the Z-direction on FDM typically is found to have the lowest tensile strength when compared to X- and Y-directions. Thus, orientation 4 containing the least height in the Z-direction is rated with a higher score. In contrast, orientation 1 has the least bottom area for SLS, which helps minimize thermal contraction by layers during the printing process and is rated with a higher score. Besides, the surface roughness data of orientations 3 and 6 obtained from the experiment for both SLS and FDM are higher (worst) than the others. However, while the value of orientation 4 in FDM is found to be the lowest R_a, implying the best surface quality, orientation 5 in SLS is found to be the best one. On the other hand, while the RMS values representing the part accuracy of orientations 4 and 5 for FDM are lower than the others (i.e. good part accuracy), orientations 1 and 4 for SLS exhibit good

Table 6.3 Comparison of the analyzed ranking list for orientation alternatives.

Ranked list		DEA technique	AHP technique		LN technique		
			Economical Type	Performance-first Type	Equal Weight	Economical Type	Performance-first Type
FDM	Alt. 1	3	2	3	3	2	3
	Alt. 2	4	5	4	4	6	4
	Alt. 3	5	4	5	6	5	6
	Alt. 4	**1**	**1**	**1**	**1**	**1**	**1**
	Alt. 5	2	3	2	2	3	2
	Alt. 6	5	6	6	5	4	5
SLS	Alt. 1	**1**	3	**1**	**1**	2	**1**
	Alt. 2	3	4	4	4	4	4
	Alt. 3	5	5	5	5	5	5
	Alt. 4	**1**	**1**	2	2	**1**	2
	Alt. 5	4	2	3	3	3	3
	Alt. 6	5	6	6	6	6	6

Remark: The bold values denote the best alternative for each category.

part accuracy with low RMS values. Also, all part orientation alternatives from SLS are found to have shrinkage.

Next, analyzed results using the integrative MCDA technique (i.e. DEA, AHP, and LN) are shown in Table 6.3. In particular, only implementing the DEA technique lacks a perspective of judgments from any decision-makers and may include a number of efficient orientation alternatives that are difficult for reaching a final decision-making (i.e. alternative 4 for FDM and alternatives 1 and 4 for SLS). In contrast, the AHP technique employed in this analysis incorporates preferences from economical decision-makers and performance-requiring decision-makers to evaluate the best ranking list (i.e. alternative 4 for FDM's economical type and performance-first type, alternative 4 for SLS's economical type, and alternative 1 for SLS's performance-first type). However, caution should be noted as the AHP technique alone requires subjective judgment at both the level of criteria evaluation and the level of alternatives with respect to criteria. In contrast, the LN technique employs AHP-based subjective weights for criteria and explicit data for the level of alternatives with respect to criteria, which allows the ranking list to be both quantitatively and qualitatively analyzed (i.e. alternative 4 for FDM's economical type, performance-first type, and equal-weight type, alternative 4 for SLS's economical type, alternative 1 for SLS's performance-first type, and alternative 1 for SLS's equal-weight type). We note that although the ranking list between AHP and LN techniques are found similar, this is not always the case since subjective judgments concerning alternative judgments from AHP are avoided.

The ranking lists obtained from the above framework can be further evaluated for trade-offs among multiple criteria to understand the validity of the proposed algorithm. Across the FDM process, alternative 4 with a perpendicular orientation for the printing platform is found to be the best orientation. Given the least build cost, build time, and little support volume, while providing the best mechanical properties and surface quality combined with moderate part accuracy, the combination of orientation 4 under simultaneous criteria consideration is shown to be the best one. Concerning the SLS process, both alternatives 4 and 1 are shown to be efficient. Whereas alternative 4 performs well when the economic preference is driven, alternative 1 comparatively performs well when the performance-first preference is motivated. This is due to the fact that orientation 4 uses the least time and the least cost in printing. On the other hand, alternative 1 is found to have the best combination of part accuracy and mechanical properties.

6.3.2 Printer Selection

We next illustrate the printer selection case study in this subsection following the key publication of Ransikarbum and Khamhong [19] to demonstrate how multiple criteria can be assessed using MCDA technique for healthcare applications. Initially, the FAHP technique is presented to assess criteria deemed important for 3D printer selection. The fuzzy logic, in particular, is incorporated using the triangular probability distribution to take uncertainty in decision-making into account. Particularly, multiple criteria/sub-criteria related to product requirements, printer characteristics, and material properties, are evaluated and synchronized in an analysis. Then, the TOPSIS technique is further employed to assess AM printer alternatives by evaluating preferences from technical experts and user groups, given a fuzzy criteria weight obtained from the FAHP earlier.

To demonstrate how MCDA can be used to suggest solutions for multiple-criteria 3DP's printer selection problem, these selective MCDA tools are briefly discussed next.

6.3.2.1 Fuzzy Analytic Hierarchy Process (FAHP)

Given that uncertain conditions exist in the real decision-making process, the integrative AHP and fuzzy logic (i.e. FAHP) can be applied. The FAHP technique, in particular, can be used for decision problems, which can be defined with (i) a set of evaluation criteria, (ii) a number of alternatives, (iii) a weighting vector, and (iv) a linguistic judgment addressing the fuzzy logic of relative importance of criteria. The FAHP is further used to assess judgments from a group of decision-makers from both expert and user groups. After the FAHP hierarchy is constructed, all the criteria at the same level of the hierarchical structure are compared against the criterion of the former level using linguistic terms for making fuzzy pairwise comparisons. We note that these linguistic terms can be used to later develop fuzzy comparison matrices. The fuzzy comparison matrix will represent fuzzy relative importance, in which the positive judgment can be treated as an inverse order of the fuzzy number of the equivalent negative judgment and vice versa.

In order to account for fuzzy comparison of the linguistic judgment, one of the popular probability distributions called triangular distribution is used, representing the worst, moderate, and the best view of decision-maker(s), respectively. Later, a triangular fuzzy number can be quantitatively interpreted using the minimum value, the most likely value, and the maximum value to account for stochastic behavior in making decisions, accordingly.

6.3.2.2 Technique for Order of Preference by Similarity to Ideal Solution (TOPSIS)

The TOPSIS technique is based on the notion that the best alternative will have the longest geometric distance from the negative ideal solution (NIS) as well as the shortest geometric distance from the positive ideal solution (PIS). In particular, relative weights obtained from the FAHP technique suggested earlier can be used in coupling with the TOPSIS technique to obtain the ranking list of 3D printer alternatives. The TOPSIS procedure is as follows. Initially, the decision matrix can be constructed, which can be further normalized using the LN technique suggested earlier. Next, the normalized decision matrix can be transformed into the weighted normalized decision matrix by multiplying it with the weights obtained earlier from the FAHP technique. The next step is to compute PIS and NIS values, respectively. Next, the separation measure can be computed, in which the positive separation measure represents the distance between the evaluated alternative and the best (i.e. ideal) condition, and the negative separation measure represents the distance between the evaluated alternative and the worst (i.e. anti-ideal) condition. Finally, the relative closeness value can be computed. In particular, the relative closeness value will be in a range between (0, 1) where 1 represents the ideal condition (i.e. the condition closest to the PIS and farthest to NIS). Accordingly, the relative closeness value will be used to obtain the ranking list for decision-makers involved in the 3D printer selection.

6.3.2.3 Illustrative Case Study for Printer Selection

To illustrate how the above MCDA tools can be applied to the 3D printer selection problem, we next deliberate the illustrative case study. Initially, 3D printer alternatives of interest can be obtained, which are illustrated as shown in Table 6.4. That is, there are five selective alternatives/printers with varied specifications. In addition, varied materials can be subjectively used for each printer alternative depending on the printer's properties. Next, results are analyzed and obtained as shown in Table 6.5, where two selective healthcare parts are used. The ranking list for 3D printers is initially analyzed when product-related criteria, material-related criteria, or printer-related criteria are independently evaluated. Trade-offs clearly exist among alternative printers, in which the best printer depends on not only how a group of decision-makers perceive multiple criteria of interest, but also on how these criteria conflict with each other. Finally, the global ranking list can be constructed, when all the relevant criteria are evaluated. The bold font, in particular, demonstrates the best (1st) option among the ranking list.

6.3.3 Part-to-Printer Scheduling

We next illustrate the case study of the part-to-printer assignment in this subsection following the key publication of Ransikarbum et al. [14] to show how multiple objectives can be modeled and assessed. In particular, the authors develop a decision-support tool for 3DP production planning involving FDM, SLA, and SLS technologies. A multi-objective optimization

Table 6.4 AM printer alternatives and material specifications.

Alternatives	AM technology	Manufacturer/size	Material type
Printer 1	SLA, Desktop XL	Germany / 100 × 75 × 100	E-Denstone
Printer 2	SLA, Ultra 3SP	Germany / 266 × 175 × 193	E-Model
Printer 3	FDM, SINDOH 3DWOX1	South Korea / 210 × 200 × 195	ABS
Printer 4	FDM, Flash forge Guider2s	China / 280 × 250 × 300	PLA
Printer 5	FDM, XYZ Da Vinci	Taiwan / 175 × 175 × 175	PET-G

Table 6.5 Analyzed results obtained from integrated FAHP and TOPSIS.

	C1 (Product) criterion		C2 (Material) criterion		C3 (Printer) criterion		Global analysis	
Alternatives	**Score (C)**	**Rank**	**Score (C)**	**Rank**	**Score (C)**	**Rank**	**Score (C)**	**Rank**
			Heart model					
Printer 1	0.7931	2	0.4013	4	0.0808	5	0.6048	4
Printer 2	**0.8147**	**1**	0.0839	5	0.3733	4	0.6160	3
Printer 3	0.7454	4	0.6164	2	0.4868	2	0.6704	2
Printer 4	0.7746	3	0.6076	3	**0.9858**	**1**	**0.7345**	**1**
Printer 5	0.1847	5	**0.8876**	**1**	0.4036	3	0.3647	5
			Jaw model					
Printer 1	0.7626	2	0.4013	4	0.0808	5	0.5572	4
Printer 2	**0.7969**	**1**	0.0839	5	0.3733	4	0.5621	3
Printer 3	0.5782	4	0.6164	2	0.4868	2	0.5680	2
Printer 4	0.6025	3	0.6076	3	**0.9858**	**1**	**0.6402**	**1**
Printer 5	0.2200	5	**0.8876**	**1**	0.4036	3	0.4172	5

approach is used to tackle varied managerial policies of the planner related to schedule activities of component batches. In addition, the AHP technique introduced earlier is applied to analyze trade-offs among conflicting criteria/objectives of interest.

In order to demonstrate how MOP/MCDA tools can be used to suggest solutions for multiple-objective 3DP's part-to-printer assignment problem, these selective tools are briefly discussed next.

6.3.3.1 Multi-objective Optimization

Multi-objective optimization (MOP) approach is concerned with modeling a system using mathematical optimization, in which more than one objective function is reflected and optimized simultaneously or sequentially. MOP has been applied in a variety of practical applications, in which trade-offs between two or more conflicting objectives exist and are evaluated [20]. In particular, a solution to MOP is called superior solution if it is feasible and optimizes all the objective functions simultaneously. However, superior solutions may not exist in most MOP problems as there are conflicts among these objective functions. Thus, the preemptive, the non-preemptive, and the epsilon constraint methods are developed to investigate the solution of

MOP models. Additionally, the Pareto front, which is the set of all Pareto efficient solutions, can also be developed to visualize trade-offs among a pair of objectives.

The preemptive approach, in particular, involves prioritizing objectives, which can be typically solved by sequential optimization. In contrast, the epsilon constraint method attempts to solve one single objective function at a time by treating other objective functions as a set of constraints. On the other hand, the non-preemptive approach uses the weighted sum method called a single-objective weighted model to solve a MOP problem. The non-preemptive approach, in particular, employs relative weights of objective functions. Thus, this solution procedure can be integrated with the F/AHP introduced earlier to analyze objective weight with the aid of objective normalization.

6.3.3.2 Illustrative Case Study for Part-to-Printer Assignment

We next discuss the case study to illustrate how the above integrated AHP and MOP tools can be applied to the part-to-printer assignment/scheduling problem. That is, the MOP model is developed with 4 managerial objectives, such that the first objective function (*Z1*) minimizes the total cost, including production and transportation costs, and the second objective function (*Z2*) maximizes the minimum of the load balance for all printers used in the production process to reflect an efficient performance with a focus on resources. In addition, the third (*Z3*) and the fourth (*Z4*) objective functions are added to minimize the total lateness, as well as the total number of unprinted parts to reflect the effective performance, respectively.

Additionally, the formulation of the mathematical approximations of two key technical requirements for 3DP of FDM, SLA, and SLS technologies are proposed: (i) the printing time approximation, and (ii) the part nesting and stacking requirement. These requirements can be adjusted to model different 3DP technologies and processes. In addition, analyzed results are obtained as illustrated in Table 6.6. Firstly, the AHP technique was used to analyze relative weights of four objective functions. Next, the LN technique was used to normalize units of each objective function to be in a range between 0 and 1. Then, the exact algorithm was later used to evaluate optimal solutions of each single objective function and multi-objective optimization problem. As shown in the table, the solution set of each single objective model highlights trade-offs among the objectives. In addition, the optimal result from the multi-objective model shows that the model tried to balance all the objectives optimally.

Table 6.6 Results of the case study based on each objective function.

Desired Criteria	Z1	Z2	Z3	Z4	Multi-Objective
Solve time (s)	0.141	0.062	0.547	0.062	0.453
Assigned printers	FDM1 SLA1 SLS1	FDM1, FDM2 SLA1, SLA2 SLS1	FDM1, FDM2 SLA1, SLA2 SLS1	FDM1, FDM2 SLA2 SLS1	FDM1, FDM2 SLA1, SLA2 SLS1
Total cost ($)	6,158	15,827	15,108	15,256	14,999
Load balance	FDM1 (58.39%) FDM2 (0%) SLA1 (38.76%) SLS2 (0%) SLS1 (3.05%)	FDM1 (58.39%) FDM2 (83.01%) SLA1 (79.08%) SLA2 (54.05%) SLS1 (5.92%)	FDM1 (58.39%) FDM2 (83.01%) SLA1 (69.62%) SLA2 (43.45%) SLS1 (4.76%)	FDM1 (58.39%) FDM2 (83.01%) SLA1 (0%) SLA2 (61.30%) SLS1 (5.92%)	FDM1 (58.39%) FDM2 (83.01%) SLA1 (69.62%) SLA2 (43.45%) SLS1 (5.92%)
Maximin load	0%	5.92%	4.76%	0%	5.92%
Total delay (days)	196	622	138	415	145
Total parts unassigned (parts)	22/22 (FDM: 7; SLA: 8; SLS: 7)	6/6 (FDM: 4; SLA: 1; SLS: 1)	8/8 (FDM: 4; SLA: 1; SLS: 3)	5/5 (FDM: 4 SLA: 0 SLS: 1)	6/6 (FDM: 4 SLA: 1 SLS: 1)

6.4 Discussion and Managerial Implication

There are a number of techniques and tools belonging to the category of MCDA approach that are used in a number of applications. Problems of analyzing the best alternative of interest under a number of multiple, conflicting criteria, in which a ranking list is desired, can be solved using a number of MCDA tools (e.g. F/AHP, TOPSIS, and DEA). These problems, in particular, can be categorized with a finite number of criteria, a finite number of alternatives, and implicit constraint lists. Besides, there are also problems that deal with finding optimal solutions under a number of multiple, conflicting objectives, which in particular involves a finite number of objectives, infinite number of possible solutions, and explicit constraint lists. The latter category belongs to the MOP classification under MCDA paradigm. Thus, it can be seen that although criteria (alternatives) and objectives (feasible solutions) can be used often interchangeably, the former is often used when a finite number of criteria and a finite number of alternatives are of interest, whereas the latter deals with a finite number of objectives and infinite number of possible solutions.

The MCDA approach, in general, has been vastly proven a successful method that can evaluate multiple, conflicting objectives/criteria in making decisions and planning for many real-world applications, inclusive of 3DP production planning. Process planning in 3DP clearly faces a similar issue, given conflicting criteria with trade-offs among objectives/criteria of interest. It is also necessary to understand the context of each particular AM technology, in which the quality of 3D printed parts highly depends on the processing printer. Thus, during the initial step of an analysis, careful consideration should be taken in selecting alternatives/evaluating possible solutions of interest as well as scrutinizing a criteria list. An expert opinion from an experienced decision-maker for 3DP technology may be acquired during this step. The multiple objectives/criteria also should be selected to reflect both consumed resources and desired outputs for 3DP production and process planning.

Finally, the choice of solution methods for 3DP production and process planning deals with the type of problems being considered. As illustrated in the DSS in this chapter, the part orientation, the printer selection, and the part-to-printer scheduling problems require different algorithms/procedures. It is worth noting that the trend of MCDA analysis is to combine and integrate various MCDA tools in order to deal with the shortcomings of each particular method alone as illustrated in the chapter.

6.5 Conclusion

3DP or AM technology has gained many communities' interests as it can provide several benefits in design flexibility, time-to-market reduction, high speed of the process, product customization, material savings, and so on. While an emphasis in the 3DP has moved toward end-use, and production parts, some issues and challenges related to conflicting objectives/criteria for decision-makers exist. Thus, obtaining a balanced solution (i.e. trade-offs among objectives/criteria) among feasible solutions becomes a must to ascertain company strategy. In this chapter, we discuss the multi-objective/criteria evaluation for 3DP, which can tackle trade-offs among key objectives/criteria relevant to various aspects of 3D printed parts and processes. The trade-offs and illustrations for selective case studies are presented through a number of selective MCDA tools for 3DP's DSS. In particular, multi-objective/criteria problems concerning part orientation, printer selection, and part-to-printer assignment for 3DP production and process planning are presented and demonstrated.

References

1 Wohlers, T. (2022). *Wohlers Report*. Wohlers Associates Inc.
2 ASTM (2012). *Standard Terminology for Additive Manufacturing Technologies*. West Conshohocken, PA: ASTM International.
3 Ransikarbum, K., Ha, S., Ma, J., and Kim, N. (2017). Multi-objective optimization analysis for part-to-printer assignment in a network of 3D fused deposition modeling. *Journal of Manufacturing Systems* 43: 35–46.
4 Zhang, Y., Harik, R., Fadel, G., and Bernard, A. (2018). A statistical method for build orientation determination in additive manufacturing. *Rapid Prototyping Journal* .
5 Di Angelo, L., Di Stefano, P., Dolatnezhadsomarin, A. et al. (2020). A reliable build orientation optimization method in additive manufacturing: the application to FDM technology. *The International Journal of Advanced Manufacturing Technology* 108 (1): 263–276.
6 Leirmo, T.S. and Martinsen, K. (2020). Deterministic part orientation in additive manufacturing using feature recognition. *Procedia CIRP* 1 (88): 405–410.
7 Ransikarbum, K., Pitakaso, R., Kim, N., and Ma, J. (2021). Multicriteria decision analysis framework for part orientation analysis in additive manufacturing. *Journal of Computational Design and Engineering* 8 (4): 1141–1157.
8 Yang, Y., Liu, B., Li, H. et al. (2022). Automatic selection system of the building orientation based on double-layer priority aggregation multi-attribute decision-making. *Journal of Intelligent Manufacturing* 1–7.
9 Peko, I., Gjeldum, N., and Bilić, B. (2018). Application of AHP, fuzzy AHP and PROMETHEE method in solving additive manufacturing process selection problem. *Tehnički vjesnik* 25 (2): 453–461.
10 Prabhu, S.R. and Ilangkumaran, M. (2019). Decision making methodology for the selection of 3D printer under fuzzy environment. *International Journal of Materials and Product Technology* 59 (3): 239–252.

11 Chen, T. and Wu, H.C. (2021). Fuzzy collaborative intelligence fuzzy analytic hierarchy process approach for selecting suitable three-dimensional printers. *Soft Computing* 25 (5): 4121–4134.

12 Raja, S. and Rajan, A.J. (2022). A decision-making model for selection of the suitable FDM machine using fuzzy TOPSIS. *Mathematical Problems in Engineering* 2022.

13 Manogharan, G., Wysk, R.A., and Harrysson, O.L. (2016). Additive manufacturing–integrated hybrid manufacturing and subtractive processes: economic model and analysis. *International Journal of Computer Integrated Manufacturing* 29 (5): 473–488.

14 Ransikarbum, K., Pitakaso, R., and Kim, N. (2020). A decision-support model for additive manufacturing scheduling using an integrative analytic hierarchy process and multi-objective optimization. *Applied Sciences* 10 (15): 5159.

15 Altekin, F.T. and Bukchin, Y. (2022). A multi-objective optimization approach for exploring the cost and makespan trade-off in additive manufacturing. *European Journal of Operational Research* 301 (1): 235–253.

16 Tafakkori, K., Tavakkoli-Moghaddam, R., and Siadat, A. (2022). Sustainable negotiation-based nesting and scheduling in additive manufacturing systems: a case study and multi-objective meta-heuristic algorithms. *Engineering Applications of Artificial Intelligence* 112: 104836.

17 Qin, Y., Qi, Q., Scott, P.J., and Jiang, X. (2019). Determination of optimal build orientation for additive manufacturing using Muirhead mean and prioritised average operators. *Journal of Intelligent Manufacturing* 30 (8): 3015–3034.

18 Justino Netto, J.M., Ragoni, I.G., Frezzatto Santos, L.E., and Silveira, Z.C. (2019). Selecting low-cost 3D printers using the AHP method: a case study. *SN Applied Sciences* (4): 1–2.

19 Ransikarbum, K. and Khamhong, P. (2021). Integrated fuzzy analytic hierarchy process and technique for order of preference by similarity to ideal solution for additive manufacturing printer selection. *Journal of Materials Engineering and Performance* 30 (9): 6481–6492.

20 Ransikarbum, K. and Mason, S.J. (2021). A bi-objective optimisation of post-disaster relief distribution and short-term network restoration using hybrid NSGA-II algorithm. *International Journal of Production Research* 1–25.

7

Role of Controlling Factors in 3D Printing

Shahriar Hashemipour[1] *and Amrid Mammeri*[2,3]

[1] *Department of Material Engineering, Iran University of Science and Technology, Tehran, Iran*
[2] *Valeo Thermal Systems, La verriere-Paris, France*
[3] *Arts et Métiers Institute of Technology, CNAM, LIFSE, HESAM University, 75013 Paris, France*

7.1 Introduction

Fused filament fabrication (FFF), classified among the array of techniques within the realm of additive manufacturing (AM), represents a pivotal method for materializing three-dimensional objects. In the FFF process, this intricate 3D architecture comes into existence as a consequence of the gradual and successive placement of layers comprised of extruded thermoplastic filament. These filaments encompass a diverse range of materials such as PLA, ABS, PP, PE, Nylon, and PEEK, each contributing its distinct characteristics to the final product. The procedure unfolds in a meticulous manner – the thermoplastic filaments are meticulously extruded in layers that mirror the x–y plane. These individual strata, assembled in a consecutive fashion along the z-direction, collaboratively culminate in the formation of a comprehensive, layer-by-layer 3D entity [1]. Integral to this process is the generation of heat by the extruder mechanism. This thermal energy facilitates the adhesion of the freshly deposited layer to its predecessor. As the latest layer is being positioned, the prior layer simultaneously undergoes a cooling phase. This juxtaposition of heat and cooling triggers a dynamic interplay wherein the substrate layers experience cycles of cooling and reheating [2, 3]. Of particular interest is the intricate relationship between adjacent filaments. The temperature profile created by this fluctuating thermal process directly influences the bonding mechanism between these neighboring filaments. This is due to the inherent cyclic nature of the temperature variations encountered by the polymer during the deposition process. In essence, the effectiveness

Industrial Strategies and Solutions for 3D Printing: Applications and Optimization,
First Edition. Edited by Hamid Reza Vanaei, Sofiane Khelladi, and Abbas Tcharkhtchi.

of the filament bonding is intricately tied to these thermal dynamics, thus underscoring the complex interplay of factors in the FFF process.

The motivation for research into diverse assessments and enhancements of components produced through FFF stems from its capacity to create intricate shapes and reduce manufacturing expenses. Notwithstanding these noted benefits, the mechanical traits of FFF-produced parts inherently exhibit deficiencies. As a result, it becomes imperative to evaluate the mechanical attributes of 3D printed materials in contrast to conventional manufacturing techniques [4, 5]. Within the FFF process, each parameter exerts its own distinct impact on both the microstructure and the bonding of filaments in the manufactured components. There exist three pivotal categories of parameters that hold sway over this process [6, 7]: (i) Material parameters encompass factors such as molecular weight, density, surface tension, thermal conductivity, heat capacity, polymer moisture content, melting temperature, crystallization temperature, and glass transition temperature. (ii) Process parameters, on the other hand, encompass variables like nozzle temperature, chamber temperature, road width, printing head speed, layer thickness, presence of air pockets, and frame angle. (iii) Machine parameters complete this trio, including attributes such as nozzle shape, nozzle temperature, print head characteristics, accuracy of positioning in the x–y plane, and accuracy of positioning in the y–z plane.

In the subsequent discussion, we will embark on a more comprehensive exploration of the significance and impact of process variables that wield control over the distinct characteristics exhibited by the 3D-printed components.

7.2 FFF Process Parameters

FFF, alternatively recognized as fused deposition modeling (FDM), made its debut in the commercial realm through Stratasys in 1992. The FFF technique predominantly employs thermoplastic filaments due to their comparatively low melting point temperatures [8]. These filaments are threaded through an extrusion nozzle that maintains an elevated temperature to induce the polymer into a viscous state. As the process involves a continual feed of solid filament in a controlled manner, the semi-liquid polymer is expelled through the nozzle, effectively generating a new layer. Simultaneously, the extruder maneuvers horizontally across the x–y plane atop the build plate, leaving behind a precisely measured layer of polymer, establishing the foundation for the impending structure. Once the entire layer has been composed, the build plate descends in the z-direction (or the printing head ascends, contingent upon the system's configuration), facilitating the deposition of a subsequent layer atop the preceding one. Consequently, the object is constructed following a bottom-up methodology [9].

The refinement of the printed components' quality is contingent upon the fine-tuning of the printing parameters, encompassing variables such as layer thickness, raster angle, raster width, air gap, and others. While the influence of these printing parameters has been explored in various studies, the majority of these investigations have been confined to the realm of conventional thermoplastics [10, 11].

An analysis of the available literature underscores the intricacy of the FFF process, wherein the caliber of the resulting components and their mechanical attributes are influenced by an extensive array of process parameters. Hence, achieving the successful production of functional engineering parts through FFF necessitates the meticulous calibration of process conditions, with the dual objectives of enhancing quality and concurrently minimizing production duration and expenses. The visual representation in Figure 7.1 outlines a comprehensive cause and effect diagram encapsulating the adjustable parameters within the FFF process, and how it is required to perform different characterizations-based studies. These parameters can be systematically categorized into six distinct groups – 3D model specifications, material properties, attributes of the FFF machine, environmental factors, printing parameter configurations, and directional considerations.

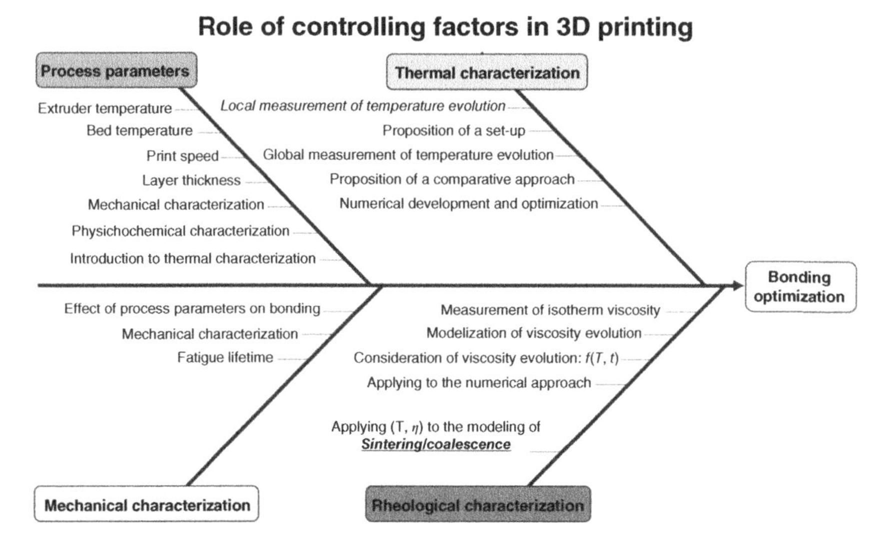

Figure 7.1 Cause and effect diagram of FFF process variables.

Notably, a subset of these parameters, particularly the printing parameters and thermal processing conditions (falling under environmental factors), can be conveniently fine-tuned within the control software to induce deliberate modifications in the ultimate properties of the manufactured component. Several primary parameters are additionally elucidated as follows:

- The magnitude of layer thickness is ascertained by the abrupt alteration in the height of either the printhead or the print bed, contingent on the specific FFF system.
- The temperature of the nozzle is indicative of the temperature at the heating block, which prompts the filament to attain a state of viscosity.
- Layer angle denotes the orientation of the layer pattern in relation to the *x*-axis as applied to the foundational part layer.
- Infill density signifies the quantity of plastic material deployed within the interior of the printer structure.

In the realm of FFF, the inherent complexity of the process is mirrored in the diverse spectrum of process parameters present across nearly all 3D-printing machines. These encompass factors such as nozzle and chamber temperatures, path width, print speed, layer thickness, the presence of air pockets, and the orientation of the frame angle. These parameters collectively contribute to the characterization of the manufactured components. Importantly, a significant portion of these parameters have a direct impact on filament adhesion, which consequently shapes the mechanical characteristics exhibited by the resulting 3D-printed objects.

Despite this multitude of parameters, researchers have sought to streamline their investigations by concentrating on key parameters that hold the potential to optimize the experimental process and yield an optimal combination of settings. This strategic focus on pivotal parameters aims to uncover an efficient configuration that balances quality, efficiency, and effectiveness in the final product. Numerous studies have delved into the exploration of the impact of process parameters on the properties of parts created using the FFF technique. For instance, the orientation in which the object is built and the angle at which its frame is positioned have been subjects of intense investigation due to their subsequent effects on the mechanical attributes of 3D-printed items. Furthermore, the influence of raster angle, determined through the analysis of infill patterns, has also been thoroughly scrutinized. These investigations collectively contribute to a deeper understanding of the intricate relationship between process parameters and the resultant properties of 3D-printed objects.

With regard to the utilized materials and the variables under investigation, it was observed that virtually all researchers made efforts to examine the influence of parameters through diverse characterization techniques such as tensile or bending tests, employing a specific parameter across various values. An abridged compilation of studies concerning a range of materials has been presented in Table 7.1.

In the realm of research on the impact of process parameters, several key studies have shed light on the intricate relationship between layer thickness and the mechanical characteristics of 3D-printed materials. Ahn et al. [22] conducted a comprehensive examination of the effect of layer thickness, specifically height, on ABS specimens. Intriguingly, their findings indicated that the material's mechanical behavior remains largely unaffected by variations in this parameter. Contrarily, Sood et al. [23] presented contrasting insights by revealing that the utilization of the smallest layer thickness value results in enhanced tensile behavior. Moreover, additional investigations have provided further insights into this matter. Other researchers have demonstrated that a reduction in layer height yields an improvement in mechanical properties.

When examining the existing body of research on process parameters, a distinct gap emerges in the study of the effects of temperature variations (including liquefier, support, and environmental temperatures) and print speed, which also plays a role in material cooling. Interestingly, previous research has primarily explored the influence of support temperature on fabricated parts' mechanical behavior. Rodriguez et al. [24] and Ahn et al. [22] have both arrived at the conclusion that support temperature does not significantly impact the mechanical properties of the final products. However, contrasting findings have been presented by Sun et al. [2] and Xiaoyong et al. [21], who have shown that support temperature does indeed exert an influence on the mechanical attributes of the printed parts. This divergence in findings highlights the complexity of the interplay between temperature variables and mechanical behavior in the context of 3D printing.

7.3 Controlling Factors as a Source of Heat Transfer

The heat source in the 3D printing process plays a pivotal role in shaping the final outcome of the manufactured objects. This heat source, often generated by extruders or lasers, serves as the driving force behind the transformation of raw materials into intricate three-dimensional structures. The intensity, distribution, and duration of heat application profoundly influence the

Table 7.1 Representation of FFF-based research in consideration of process parameters.

Material	Variable parameters	Mechanical properties	
PLA	Layer thickness	Shear stress	[12]
	Infill density		
	Post-processing heat treatment at $T = 100\,°C$		
ABS	Five building orientations on x and y axis	Impact strength	[13]
ABS	Two orientations (at x axis – 0, 90°)	Tensile strength	[14]
	Y axis orientation (0, 30°)		
ABS	Raster orientation	Tensile strength	[15]
	Number of layers (1–35)	Elastic modulus	
		Elongation at break	
ABS	Five-layers orientation	Tensile strength	[16]
	(45/−45, 0, 45, 90, 45/0)	Modulus	
		Impact resistance	
ABS	Raster orientation	Tensile strength	[17]
	Air gap	Compressive strength	
	model temperature	Comparison of results with injection molding	
PLA	Effect of process parameters on bonding	Tensile strength	[18]
PLA	Deposition orientation	Tensile strength	[19]
	Layer thickness	Flexural strength	
	Raster variation	Impact strength	
PEEK	Layer thickness	Tensile	[20]
ABS	Raster angle	Compression	
		Bending strength	
PEEK	Temperature variation (bed and environment)	Tensile strength	[21]
ABS	Temperature profile	Three-point bending test	[2]
	Temperature variation with part building	Thermal analysis	

material's behavior during each layer's deposition and subsequent solidification. One of the key effects of the heat source is the material's transition from a solid to a molten or semi-molten state. As the heat source interacts with the raw material – typically thermoplastic filaments or powders – it raises their temperature above their melting or softening point. This controlled heating process enables the material to become more pliable and amenable to shaping, facilitating the layer-by-layer construction of the object.

Additionally, the heat source impacts the bonding between layers and the overall structural integrity of the printed part. Proper adhesion between adjacent layers is crucial to ensure the mechanical strength of the final object. In FFF or FDM, for example, the heat from the extruder melts the material just enough to create a strong bond between layers while minimizing distortion. On the other hand, in selective laser sintering (SLS) or stereolithography (SLA), lasers or other energy sources selectively solidify, or fuse powdered or liquid materials, creating tightly interconnected structures layer-by-layer.

Controlling the heat source also affects the cooling rate of the material after deposition. Rapid cooling can lead to thermal stresses and distortions within the printed object. Conversely, controlled cooling can promote crystallinity and enhance material properties. Moreover, the heat source's precision and accuracy influence the resolution and level of detail achievable in the final printed object. Highly controlled and localized heat application enables the creation of intricate features and finer layers, resulting in smoother surfaces and higher precision. The heat source acts as the catalyst for the transformative process in 3D printing, affecting material behavior, layer adhesion, mechanical properties, cooling dynamics, and overall printing precision. The interplay between heat, material, and design parameters necessitates careful calibration to achieve the desired quality and functionality of the final printed object.

7.4 Impact of Controlling Factors on Mechanical Features of 3D-Printed Parts

The mechanical characteristics of 3D-printed parts are profoundly influenced by a multitude of controlling factors throughout the AM process. These factors encompass a wide spectrum, ranging from design parameters to material properties and printing conditions. The interplay of these factors intricately shapes the final mechanical attributes of the printed objects, dictating their strength, durability, flexibility, and other essential properties.

Design parameters, such as geometry, infill density, and orientation, have a substantial impact on the mechanical behavior of 3D-printed parts. The chosen geometry and infill pattern determine the internal structure, which in turn affects factors like load distribution and stress propagation. Design orientation influences the anisotropic behavior of printed parts, where mechanical properties can significantly differ in different directions. Material selection is another critical factor. Different types of materials, such as polymers, metals, ceramics, and composites, exhibit diverse mechanical characteristics. The material's intrinsic properties, such as tensile strength, elasticity, and thermal expansion, directly influence the final part's behavior under various stresses.

Printing parameters encompass variables like layer height, print speed, temperature, and cooling rate. The layer height influences surface roughness and layer adhesion, while print speed and temperature control the material's deposition and bonding process. Effective cooling strategies are essential to prevent warping and thermal stress accumulation during the printing process. Layer bonding is a pivotal aspect affecting mechanical properties. Strong inter-layer adhesion is crucial for achieving uniform and reliable mechanical strength. The adequacy of layer bonding depends on factors like temperature management, material compatibility, and extrusion precision. Post-processing techniques, including annealing, curing, and polishing, further refine mechanical properties. These processes can enhance the material's crystallinity, reduce internal stresses, and improve surface finish, consequently affecting mechanical behavior. To optimize mechanical characteristics, a holistic approach is required. A deep understanding of how these controlling factors interact is imperative for tailoring 3D-printed parts to meet specific performance requirements. As AM continues to evolve, comprehensive exploration of these factors and their impact on mechanical properties is essential for the production of high-quality, functional, and reliable 3D-printed components.

7.5 Role of Controlling Factors on Interfacial Bonding of 3D-Printed Parts

The interfacial bonding within 3D-printed parts is profoundly influenced by a multitude of controlling factors that govern the interactions between successive layers of material. These factors encompass a complex interplay of material properties, printing parameters, and environmental conditions, collectively shaping the strength, integrity, and reliability of the bond between adjacent layers. Material properties play a critical role in determining interfacial bonding. The rheological behavior and melt characteristics of the material directly impact its ability to fuse with neighboring layers. Materials with suitable viscosity and thermal properties are more likely to achieve strong inter-layer adhesion, resulting in robust final parts.

Printing parameters, such as nozzle temperature, print speed, and layer height, significantly affect interfacial bonding. The nozzle temperature governs the material's melting and flow behavior, impacting its ability to fuse with the underlying layer. Optimal nozzle temperature ensures proper fusion without causing distortion or warping. Print speed and layer height influence the deposition process, determining the layer's thickness and uniformity, which in turn affects the strength of the bond. Environmental conditions, particularly cooling rates, are instrumental in achieving well bonded layers. Rapid cooling can lead to inadequate adhesion, as insufficient time is provided for layers to properly fuse. Controlled cooling mechanisms, such as heated build chambers or controlled ambient temperatures, facilitate gradual cooling, allowing for improved inter-layer bonding.

Surface preparation and post-processing techniques also play a role in enhancing interfacial bonding. Proper surface treatments, such as texturing or chemical treatments, can promote mechanical interlocking between layers. Post-processing methods like annealing or curing can help relieve internal stresses, and enhance layer cohesion, and overall interfacial strength. The role of each controlling factor is intricate, and their synergistic effects determine the quality of inter-layer bonding. Achieving optimal interfacial bonding requires a thorough understanding of these factors and their interactions. Tailoring material properties, adjusting printing parameters, and employing suitable post-processing methods are all pivotal strategies for achieving strong and reliable bonds between layers in 3D-printed parts.

7.6 Role of Controlling Factors on Optimization of 3D-Printed Parts

The optimization of the 3D printing process is a multifaceted endeavor aimed at enhancing the overall quality, efficiency, and reliability of AM. This process involves a meticulous assessment and adjustment of various parameters, spanning material selection and preparation, design considerations, printing parameters, and post-processing techniques. Starting with material optimization, selecting the appropriate material for a specific application is critical. Material properties such as strength, flexibility, thermal conductivity, and chemical resistance must align with the intended use of the printed object. Furthermore, optimizing material formulations or incorporating additives can enhance printability and end-use performance.

Design optimization involves tailoring the 3D model to suit the intended purpose while taking advantage of AM's unique capabilities. This includes considerations such as geometrical complexity, infill patterns, and support structures. Design software and generative design tools play a pivotal role in optimizing geometries for enhanced functionality and reduced material usage. Fine-tuning printing parameters is a central aspect of process optimization. This entails adjusting factors like layer height, print speed, nozzle temperature, and cooling settings to achieve the desired balance between speed, accuracy, and part quality. An iterative approach, guided by experimentation and data analysis, is often used to refine these parameters.

Post-processing methods further contribute to optimization. Techniques such as annealing, curing, or surface finishing can refine mechanical properties, reduce residual stresses, and improve surface quality. Implementing effective post-processing strategies ensures that the printed parts meet or exceed performance expectations. Process optimization also involves harnessing advanced technologies like real-time monitoring and closed-loop control systems. These technologies enable continuous feedback during the printing process, allowing for immediate adjustments to maintain desired outcomes. Ultimately, the goal of optimization is to achieve consistent and reproducible results while minimizing material waste, production time, and costs. Through a holistic approach that addresses material, design, printing parameters, and post-processing techniques, the 3D printing process can be fine-tuned to deliver high-quality, functional, and reliable products across various industries and applications.

Layer thickness holds significant importance due to its diverse implications. It affects various aspects such as the surface texture of the 3D printed components, the production time and cost of the part, and notably, the mechanical characteristics. A larger layer thickness contributes to decreased adhesion, resulting in increased voids that manifest as porosity. This, in turn, raises the potential for cracks to develop within the printed part.

In an illustrative case, Anitha et al. [25] embarked on a thorough investigation into the impact of vital FDM process parameters on the surface roughness of ABS prototypes. To conduct this study, they employed Taguchi's design matrix, along with the signal-to-noise ratio (S/N) and analysis of variance (ANOVA) techniques. Their focus encompassed three critical process parameters – layer thickness, road width, and deposition speed. Through their study, several significant findings emerged. The paramount factor influencing surface roughness was identified as layer thickness. This parameter's pivotal role in determining the degree of roughness stems from its inherent connection to the resolution of the printed layers. Specifically, the resolution is influenced by the air gap between adjacent layers. Notably, the researchers highlighted that higher layer thickness resulted in reduced surface roughness. This correlation implies that a greater layer thickness corresponds to a smoother surface finish, as opposed to the effects of road width and speed. Furthermore, the study uncovered an inverse relationship between layer thickness and surface roughness. This implies that as layer thickness increases, the roughness of the surface diminishes. This correlation is a direct outcome of the resolution mechanism linked to layer thickness, where a larger gap between layers contributes to smoother surfaces. Ultimately, this investigation by Anitha *et al.* offers valuable insights into the intricacies of FDM process parameters and their consequential impact on the surface quality of 3D-printed ABS prototypes.

3D-printed components diverge from traditional manufacturing in that their internal structure is not uniformly solid. Rather, the AM process employs various patterns to construct the interior surfaces of these parts. This innovative approach yields multiple benefits, primarily a substantial reduction in costs due to minimized material consumption and printing time. However, this efficiency-driven technique does entail a moderate compromise in strength. The density of the interior pattern, known as the infill percentage, governs the trade-off between these factors. This pivotal printing parameter, the infill percentage, exercises a significant influence over the ultimate tensile strength of the printed components. Notably, an unequivocal correlation has been established between the infill percentage and the material's strength, where a higher infill density translates to increased overall robustness. Interestingly, the ramifications of infill density extend beyond strength considerations alone. It has been uncovered that infill density also bears an impact on the surface finish and dimensional accuracy of the resultant parts. Empirical evidence has demonstrated that opting for a higher infill density yields superior outcomes in terms of both surface texture and precise adherence to specified dimensions. In essence, the choice of infill percentage entails a thoughtful assessment of trade-offs between material strength, production efficiency, and final part aesthetics and accuracy. Striking the right balance in infill density ensures that 3D-printed objects fulfill the intended functional requirements while optimizing resource utilization and maintaining desirable surface quality and dimensional fidelity [26, 27].

As previously mentioned, there are some parameters associated with heat transfer and temperature variation during the fabrication process such as print speed, nozzle temperature, and platform temperature.

The print speed in the 3D printing process is a crucial parameter that exerts a significant influence on the overall optimization of AM. The rate at which material is deposited during printing directly affects multiple aspects of the final part's quality, efficiency, and properties. Adjusting print speed strategically plays a pivotal role in achieving a balance between production time, part accuracy, and mechanical characteristics. Print speed optimization has a direct impact on production time, as faster printing speeds expedite the construction of each layer. However, the trade-off lies in the potential compromise of part accuracy and surface quality. Rapid printing may lead to increased vibrations and forces that can introduce dimensional inaccuracies and surface imperfections [28].

Furthermore, the mechanical attributes of the printed part are intricately linked to the chosen print speed. Higher print speeds can introduce higher internal stresses due to increased material deposition rates and cooling rates, potentially resulting in reduced mechanical strength and durability. Conversely, slower print speeds offer a more controlled deposition process, allowing for enhanced layer adhesion and reduced internal stress accumulation, thereby yielding improved mechanical properties. The effect of print speed extends to material behavior as well. Faster deposition rates can lead to higher temperature differentials within the material, affecting its melting and solidification behavior. This, in turn, can influence properties such as material crystallinity and bonding between layers. Balancing these considerations requires a thorough understanding of the material being used, the specific 3D printing technology employed, and the desired part characteristics. An iterative approach involving experimentation and analysis is often employed to determine the optimal print speed that aligns with the intended balance between production time, part accuracy, and mechanical performance. Ultimately, optimizing print speed plays a vital role in achieving high-quality 3D-printed components that meet both functional and aesthetic requirements, contributing to the overall success of the AM process [29].

Nozzle temperature and platform temperature wield considerable influence over the optimization of the 3D printing process, significantly impacting various aspects of print quality, material behavior, and part adhesion. These two parameters play distinct roles in ensuring the successful execution of AM, with nozzle temperature primarily governing material extrusion and platform temperature affecting the material's interaction with the build surface. Nozzle temperature, the heat applied to the extruder, is a pivotal factor that directly influences material viscosity, flow behavior, and layer bonding. Optimal nozzle temperature ensures proper material extrusion, with excessive heat potentially leading to over-extrusion and poor layer adhesion. In contrast, insufficient nozzle temperature might hinder material flow and adhesion. The choice of nozzle temperature depends on the specific material being used and its melting characteristics. Striking the right balance facilitates uniform material deposition, precise layer bonding, and accurate part geometry. Platform temperature, on the other hand, influences the initial stages of part adhesion to the build surface. Elevated platform temperature helps mitigate warping and detachment of the printed part during the printing process, especially for materials prone to shrinkage as they cool. Proper platform temperature fosters secure adhesion and promotes even layer deposition, reducing the likelihood of defects and layer misalignment. However, excessively high platform temperatures can lead to material deformation and distortion [28].

The interplay between these two temperature parameters and their effects on optimization underscores the intricate nature of 3D printing. Achieving the right balance between nozzle temperature and platform temperature is crucial for producing parts with consistent quality, structural integrity, and dimensional accuracy. This balance is material-dependent and involves experimentation and fine-tuning to ensure that the chosen temperatures facilitate successful material deposition, adhesion, and cooling, ultimately resulting in well crafted 3D-printed components.

7.7 Summary and Outlook

In 3D printing, the role of controlling factors is undeniably paramount, exerting a profound impact on the quality, functionality, and efficiency of the AM process. The intricate interplay between various parameters, spanning material properties, design considerations, printing settings, and post-processing techniques, shapes the final outcome of the fabricated objects. The optimization of these factors aligns with the quest for producing components that not only meet but exceed performance expectations. Material properties stand as a cornerstone in this process, with the choice of material dictating mechanical attributes, thermal behavior, and material flow during deposition. Design parameters influence both the aesthetics and functionality of printed parts, determining internal structures, infill patterns, and overall geometries. Strategic adjustments of printing parameters, including layer thickness, print speed, and temperature, fine-tune the deposition process, influencing layer adhesion, surface finish, and mechanical strength. Moreover, post-processing methods offer avenues for enhancing properties and aesthetics, further underlining the holistic nature of optimization.

The collaboration between these controlling factors leads to a harmonious orchestration of the AM process, enabling the production of parts with tailored attributes. This synthesis aligns with the broader goal of 3D printing – crafting components with exceptional precision, customized designs, and enhanced performance characteristics. As the field continues to evolve, the understanding and manipulation of these controlling factors will remain pivotal, paving the way for innovation across industries and ushering in a new era of advanced manufacturing possibilities.

References

1 Chennakesava, P. and Narayan, Y.S. (2014). Fused deposition modeling-insights. In: *Proceedings of the International Conference on Advances in Design and Manufacturing ICAD&M*.

2 Sun, Q., Rizvi, G.M., Bellehumuer, C.T., and Gu, P. (2008). Effect of processing conditions on the bonding quality of FDM polymer filaments. *Rapid Prototyping Journal* 14 (2): 72–80.

3 Bellehumeur, C., Bisaria, M., and Vlachopoulos, J. (1996). An experimental study and model assessment of polymer sintering. *Polymer Engineering & Science* 36 (17): 2198–2207.

4 Caminero, M.Á., Chacon, J.M., Garcia-Plaza, E. et al. (2019). Additive manufacturing of PLA-based composites using fused filament fabrication: effect of graphene nanoplatelet reinforcement on mechanical properties, dimensional accuracy and texture. *Polymers* 11 (5): 799.

5 Depuydt, D., Balthazar, M., Hendrickx, K. et al. (2019). Production and characterization of bamboo and flax fiber reinforced polylactic acid filaments for fused deposition modeling (FDM). *Polymer Composites* 40 (5): 1951–1963.

6 Cuiffo, M.A., Synder, J., Elliott, A.M. et al. (2017). Impact of the fused deposition (FDM) printing process on polylactic acid (PLA) chemistry and structure. *Applied Sciences* 7 (6): 579.

7 Chacón, J., Caminero, M.A., Garcia-Plaza, E. et al. (2017). Additive manufacturing of PLA structures using fused deposition modelling: effect of process parameters on mechanical properties and their optimal selection. *Materials & Design* 124: 143–157.

8 Hsiao, B.S. and Sauer, B.B. (1993). Glass transition, crystallization, and morphology relationships in miscible poly (aryl ether ketones) and poly(ether imide) blends. *Journal of Polymer Science Part B: Polymer Physics* 31 (8): 901–915.

9 Solomon, I.J., Sevvel, P., and Gunasekaran, J. (2021). A review on the various processing parameters in FDM. *Materials Today: Proceedings* 37: 509–514.

10 Letcher, T. and Waytashek, M. (2014). Material property testing of 3D-printed specimen in PLA on an entry-level 3D printer. In: *ASME International Mechanical Engineering Congress and Exposition*. American Society of Mechanical Engineers.

11 Ziemian, C., Sharma, M., and Ziemian, S. (2012). Anisotropic mechanical properties of ABS parts fabricated by fused deposition modelling. *Mechanical Engineering* 23: 159–180.

12 Torres, J., Cotelo, J., Karl, J., and Gordon, A.P. (2015). Mechanical property optimization of FDM PLA in shear with multiple objectives. *JOM* 67 (5): 1183–1193.

13 Górski, F., Kuczko, W., and Wichniarek, R. (2014). Impact strength of ABS parts manufactured using fused deposition modeling technology. *Archives of Mechanical Technology and Automation* 31 (1): 3–12.

14 Górski, F., Wichniarek, R., Kuczko, W., and Andrzejewski, J. (2015). Experimental determination of critical orientation of ABS parts manufactured using fused deposition modelling technology. *Journal of Machine Engineering* 15 (4): 121–132.

15 Letcher, T., Rankouhi, B., and Javadpour, S. (2015). Experimental study of mechanical properties of additively manufactured ABS plastic as a function of layer parameters. In: *ASME 2015 International Mechanical Engineering Congress and Exposition*. American Society of Mechanical Engineers Digital Collection.

16 Es-Said, O., Foyos, J., Noorani, R. et al. (2000). Effect of layer orientation on mechanical properties of rapid prototyped samples. *Materials and Manufacturing Processes* 15 (1): 107–122.

17 Ahn, S.H., Odell, D., Roundy, S. et al. (2002). Anisotropic material properties of fused deposition modeling ABS. *Rapid Prototyping Journal* .

18 Li, H., Wang, T., Sun, J. et al. (2018). The effect of process parameters in fused deposition modelling on bonding degree and mechanical properties. *Rapid Prototyping Journal* 24 (1): 13.

19 Liu, X., Zhang, M., Li, S. et al. (2017). Mechanical property parametric appraisal of fused deposition modeling parts based on the gray Taguchi method. *The International Journal of Advanced Manufacturing Technology* 89 (5–8): 2387–2397.

20 Wu, W., Geng, P., Li, G. et al. (2015). Influence of layer thickness and raster angle on the mechanical properties of 3D-printed PEEK and a comparative mechanical study between PEEK and ABS. *Materials* 8 (9): 5834–5846.

21 Xiaoyong, S., Liangcheng, C., Honglin, M. et al. (2017). Experimental analysis of high temperature PEEK materials on 3D printing test. In: *2017 9th International Conference on Measuring Technology and Mechatronics Automation (ICMTMA)*. IEEE.

22 Ahn, S.H., Montero, M., Wright, P. et al. (2002). Anisotropic material properties of fused deposition modeling ABS. *Rapid Prototyping Journal* 8 (4): 248–257.

23 Sood, A.K., Ohdar, R.K., and Mahapatra, S.S. (2010). Parametric appraisal of mechanical property of fused deposition modelling processed parts. *Materials & Design* 31 (1): 287–295.

24 Rodríguez, J.F., Thomas, J.P., and Renaud, J.E. (2003). Mechanical behavior of acrylonitrile butadiene styrene fused deposition materials modeling. *Rapid Prototyping Journal* 9 (4): 219–230.

25 Anitha, R., Arunachalam, S., and Radhakrishnan, P. (2001). Critical parameters influencing the quality of prototypes in fused deposition modelling. *Journal of Materials Processing Technology* 118 (1–3): 385–388.

26 Abeykoon, C., Sri-Amphorn, P., and Fernando, A. (2020). Optimization of fused deposition modeling parameters for improved PLA and ABS 3D printed structures. *International Journal of Lightweight Materials and Manufacture* 3 (3): 284–297.

27 Jackson, B., Fouladi, K., and Eslami, B. (2022). Multi-parameter optimization of 3D printing condition for enhanced quality and strength. *Polymers* 14 (8): 1586.

28 Deng, X., Zeng, Z., Peng, B. et al. (2018). Mechanical properties optimization of poly-ether-ether-ketone via fused deposition modeling. *Materials* 11 (2): 216.

29 Attoye, S., Malekipour, E., and El-Mounayri, H. (2019). Correlation between process parameters and mechanical properties in parts printed by the fused deposition modeling process. In: *Mechanics of Additive and Advanced Manufacturing, Volume 8: Proceedings of the 2018 Annual Conference on Experimental and Applied Mechanics*. Springer.

8

Physico-chemical Features of 3D-printed Parts

Wuzhen Huang and Yi Xiong

School of System Design and Intelligent Manufacturing, Southern University of Science and Technology, Shenzhen, PR China

8.1 Introduction

Additive manufacturing (AM), also known as three-dimensional (3D) printing technology, allows the fabrication of high-performance parts with complex designs [1]. In contrast to other manufacturing technologies, AM creates 3D physical objects from digital models by adding materials in a layer-wise manner. The past three decades have witnessed a fast development of various AM techniques, including material extrusion, vat polymerization, powder bed fusion, and directed energy deposition. Fused deposition modeling, also known as fused filament fabrication (FFF), is one of the most widely used methods, featuring its highly cost-effective and easy-to-use methods. The advance of this technology has shifted it from a prototyping tool toward a viable production method for high-quality industrial-grade products. In this context, building an understanding of the physicochemical characteristics of fabricated parts and their relationships with process parameters becomes critical for high-end applications such as aerospace engineering and biomedical engineering.

This chapter aims to organize related knowledge surrounding physico-chemical characteristics of parts fabricated by fused deposition modeling. The remainder of this chapter is organized as follows. Section 8.2 introduces the basics of FFF and commonly used feedstock materials. Then, the physicochemical characterization of 3D-printed parts and as-received polymer materials are presented. Finally, the effect of phase change on the quality of 3D-printed parts is covered.

Industrial Strategies and Solutions for 3D Printing: Applications and Optimization,
First Edition. Edited by Hamid Reza Vanaei, Sofiane Khelladi, and Abbas Tcharkhtchi.

8.2 Fused Filament Fabrication

In the FFF process, the filament is melted and extruded in 2D tool paths layer-by-layer to build up the desired object. Several basic principles that are common to the FFF printing system [2, 3]. First, the system needs to load a continuous filament material. During a state-of-the-art FFF process, a solid thermoplastic filament is hauled off into a hot die by two counter-rotating driving wheels (bearing and gear), as shown in Figure 8.1, thus providing a mechanism for generating an input pressure for the nozzle. The spooled filaments, typically prepared by extrusion of any thermoplastic polymer, are transported through a moving deposition unit onto a heated build platform, resulting in a layer-by-layer fabrication of the structural element according to CAD-defined layer contours. Depending on the position of the extruder and nozzle during the material loading process, FFF printing system can be divided into two main types: direct drive Figure 8.1a and indirect drive systems (also called Bowden system) Figure 8.1b. The main difference between the two types is the distance between the motor drive gear and heater. The motor drive gear of the direct-extrusion printer is installed directly above the heater while that of the Bowden extrusion printer is installed away from the heater, and the filament is pulled through a long Bowden tube into the heater for printing. Although the Bowden printer has better accuracy, it has more requirements for the flexibility and stiffness of the filament than the direct-extrusion printer owing to the Bowden tube [4]. Therefore, to ensure successful 3D printing, the selected polymer and polymer combinations should have good printability.

Second, liquification, extrusion, and solidification of the material are the three states in the FFF of polymer materials. To be extruded through the

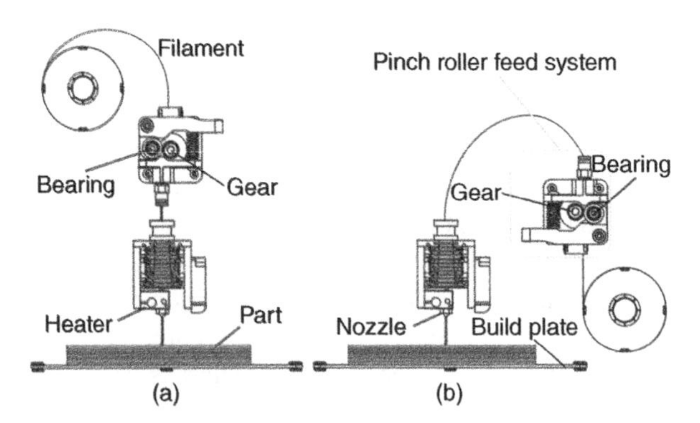

Figure 8.1 The diagram of fused filament fabrication process: (a) direct extrusion (b) Bowden extrusion.

nozzle, the filament is fed into the liquefier under motor drive gear and heated above the melting temperature of semicrystalline thermoplastic filaments. After leaving the nozzle, the extruded material is deposited layer by layer onto a building platform or a previous layer in the horizontal plane at a certain speed and pressure; the deposited melt cools down and resolidifies.

Third, material bonding to form a coherent solid structure also plays an important role in FFF processing. For the FFF printing system, there must be sufficient residual heat energy to activate the surfaces of the adjacent regions, causing bonding. In this case, we visualize the process in terms of energy supplied to the material by the extrusion head. If there is insufficient energy, the regions may adhere, but there would be a distinct boundary between new and previously deposited materials. This can represent a fractured surface where the materials can be easily separated. Too much energy may cause the previously deposited material to flow, which in turn may result in a poorly defined part. Once the material has been extruded, it must solidify and bond with adjacent material. Therefore, many scholars have defined the relevant governing equations to describe the thermal process of working on the extruded road.

Over the other manufacturing process, the process can be used to construct complex geometries that cannot be produced by conventional manufacturing processes. Besides, the efficiency of the FFF printing process is relatively high with minimal post-processing. The FFF technology is commonly used for prototyping and rapid manufacturing. A suitable FFF printing system can also be applied to multiple different materials, and it also has a large selection of materials, including many common thermoplastics, so it is also applied in manufacturing within other sectors, including m aerospace, automotive, construction, electronics, and energy industries.

8.3 Different Types of Applicable Materials in FFF

Traditional polymers can be divided into two broad categories: thermosetting polymers and thermoplastic polymers. Thermosetting polymers are polymers that typically start as viscous liquids, and, when subjected to heat or pressure, harden to form a solid. Thermoplastic polymers, the subject of this chapter, soften and melt to be shaped when heated and then harden once cooled. Based on the working principle of FFF, the materials heated at the nozzle need to be extruded by the pressure of the unmelted polymers, so the printed polymers required for FFF printing have a lower viscosity and higher mechanical strength. It should also have a relatively low glass transition temperature and melting temperature. So far, there are a variety

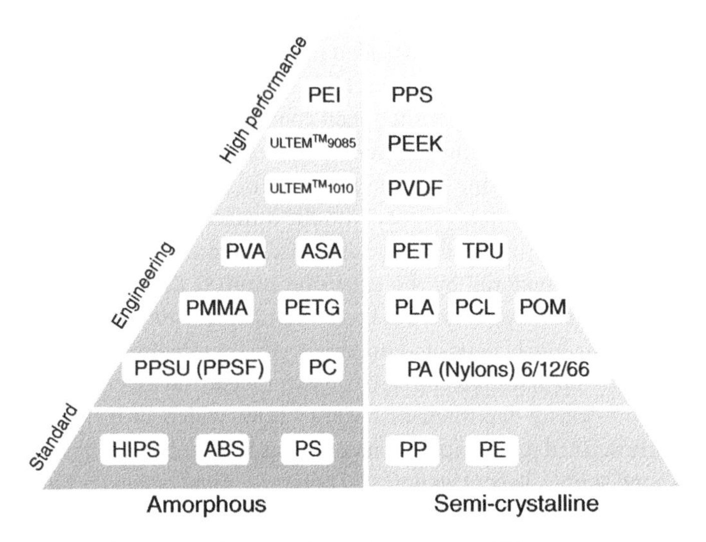

Figure 8.2 An overview of polymer filaments for FFF.

of pure polymer materials used for FFF 3D printing technology, most of which are thermoplastic polymers.

The widespread use of thermoplastic polymers has revolutionized our lives. There is a wide range from commodity standard plastics up to high-performance polymers that can perform in very demanding durable applications. According to the properties and uses, we can divide polymer species into three major categories: standard polymers (commodity polymers), engineering polymers, and high-performance polymers (HPP) (Figure 8.2).

Standard Polymers: Standard polymers are polymers that are used in high volumes for a diversity of applications. Polyolefins including polypropylene (PP), polyethylene (PE), and polybutylene (PB) are polymers that have been applied to the FFF process. They are readily available in large quantities and have reasonable performance for the applications in which they are used. PE, for example, is commonly used to make high-volume items such as plastic bags and packaging materials, applications where high temperature or extreme strength is not a critical factor.

Engineering Polymers: In contrast, engineering plastics are a distinct step up from commodity polymers and offer a superior range of properties, particularly in mechanical and thermal performance, but with superior performance comes a higher price. These kinds of polymers have replaced traditional materials in the automotive, textile, food, and electronics industries. For example, the polyamide family (Nylons) has a wide range of applications, from textiles for clothing to food and beverage containers, and in engineering applications like domestic appliances and automotive components.

High-Performance Polymers: High-performance polymers are located at the top of the polymer pyramid. They are characterized by their high mechanical performance and excellent thermal and chemical resistance. They can retain their desirable properties even when subjected to harsh conditions and can serve as a lightweight alternative for applications that involve high operating temperatures. However, they are typically much more difficult to produce and based on more complex monomers and are generally more expensive as a result. Polyaryletherketones (PAEKs), polyphenylene sulfide (PPS), and polyetherimide (PEI) belong to this class of polymers and have been studied for FFF. Due to their high melting temperatures, high-temperature FFF systems are needed to process these polymers.

Polymer filament used for FFF can also be divided into two types according to its crystallinity, namely amorphous polymers, and semi-crystalline polymers. The molecular arrangement of amorphous polymers is relatively irregular, and the molecular chains are also randomly arranged; Semi-crystalline polymers present a crystalline or amorphous state under different fabrication conditions. The regular arrangement region is called the crystal region, and the disordered arrangement region is called the amorphous region.

8.3.1 Classification of Polymers

8.3.1.1 Amorphous Polymers

8.3.1.1.1 Acrylonitrile Butadiene Styrene (ABS) Acrylonitrile butadiene styrene (ABS) is made by various acrylonitrile blends and copolymers, butadiene-containing polymers, and styrene. ABS was introduced in the 1950s as a stricter alternative to styrene–acrylonitrile (SAN) copolymers. The material's color is white; however, the oxidation of these polymers may result in a yellowing coloration. The glass transition temperature of ABS was 105 °C (221 °F), making it perfect for use within relatively safe machines that are easy to function. The low melting point makes ABS material easy to serve with a desktop 3D printer, but it has some great stuff possessions. ABS is a ductile material with wear resistance and heat tolerance. ABS is lightweight with good impact strength; nevertheless, it is really abrasion-resistant and inexpensive. Furthermore, ABS polymers defy many chemical formulations.

8.3.1.1.2 Polystyrene (PS) Polystyrene (PS) is an amorphous, glassy polymer that is generally rigid and relatively inexpensive. Unfilled PS has a sparkle appearance and is often referred to as crystal PS or general-purpose polystyrene (GPPS). Although it is similar to ABS, its positive qualities such as its good impact resistance and ease of finishing make it a strong contender against the other plastics out there.

8.3.1.1.3 High Impact Polystyrene (HIPS) High impact polystyrene (HIPS) is 3D printing material with an excellent printing quality. It is produced by adding rubber or butadiene copolymer. Therefore, this increases the toughness and impact strength of the polymer. It possesses the properties of good impact resistance and low water absorption. It can also be used as the breakaway support material for ABS, ASA, and PC. HIPS plastic material is specified for low-strength structural applications wherein impact resistance, machinability, and low cost are prerequisites. It is also more frequently used for machining pre-production prototypes because of its excellent dimensional stability and easy-to-fabricate paint and glue capabilities.

8.3.1.1.4 Polyvinyl Alcohol (PVA) PVA is an abbreviation for polyvinyl alcohol, a water-soluble material. It is often used with multi-extruder FDM 3D printers as a support material. In 3D printing, PVA's water solubility makes it excellent support for models that include intricate detailing cavities or involve overhanging features. Not only is it water-soluble, but it is also odorless, colorless, soft, biodegradable, and non-toxic at low concentrations.

8.3.1.1.5 Acrylonitrile Styrene Acrylate (ASA) ASA filament is an all-purpose prototyping material. It is the perfect all-purpose 3D printing thermoplastic, suitable for many different applications. It has a similar chemical makeup to ABS plastic but with better mechanical properties, superior aesthetics, and it is UV resistance. ASA is especially well-suited for outdoor, commercial, and infrastructure-related production parts. ASA does not degrade with prolonged outdoor use. 3D parts printed with ASA are accurate, stable, and very durable.

8.3.1.1.6 Polymethyl Methacrylate (PMMA) Polymethyl methacrylate (PMMA) is a strong, lightweight, and transparent thermoplastic. It is commonly known as acrylic and is marketed commercially as shatterproof.

PMMA filament has excellent impact strength that is significantly higher than glass. This is an ideal material when you need a hard object, with little flex, excellent finish quality, and clarity. In addition, it has half the density of glass with comparable transparency and UV absorption properties. PMMA responds very well to post-print finishing, such as sanding, drilling, or engraving. And PMMA lends itself perfectly to the process of lost wax casting due to its ability to burn away cleanly without residue. It also has pure resin quality. The natural translucency and transparency of PMMA filament make it the ideal solution where a clear part or object is the desired result. Applications where the 3D print is one vital stage in the process of creating the final part.

8.3.1.1.7 Polyethylene Terephthalate Glycol (PETG): Polyethylene terephthalate glycol, commonly known as PETG or PET-G, is a rigid co-polyester thermoplastic material that provides significant chemical resistance against many industrial fluids, durability, good toughness, wear resistance, and excellent formability for manufacturing. PETG is an amorphous plastic, which means that the molecular chains are not aligned or ordered in the material. PETG is made from PET and glycol. PET supplies the rigidity and glycol is used to lower the brittle effect of PET. PETG is produced in the same way as PVC and has properties somewhat similar to this one, its maximum temperature and its thermoformability in particular.

8.3.1.1.8 Polyphenylsulfone (PPSU, PPSF) Polyphenylsulfone (PPSU, PPSF) material has the greatest heat and chemical resistance. It is ideal for aerospace, automotive, and medical applications. PPSF parts are not only mechanically superior but also dimensionally accurate, to better predict end-product performance. Users can also sterilize PPSF via steam autoclave, EtO sterilization, plasma sterilization, chemical sterilization, and radiation. PPSF gives designers the ability to manufacture parts directly from digital files that are ideal for conceptual modeling, functional prototyping, manufacturing tools, and production parts.

8.3.1.1.9 Polycarbonate (PC) Polycarbonate (PC) is available in white and is compatible with both breakaway and soluble support materials. PC is characterized by its high strength and impact resistance, coupled with dimensional stability and heat resistance up to 110 °C. Its stiffness and stability mean the parts you print retain their shape and dimensional accuracy. High impact strength makes this a tough material, suitable for high-durability applications like functional prototyping and manufacturing tooling. These attributes make it a good choice for 3D-printed prototypes, parts, and tools that demand higher material properties than ABS or ASA. It is a FFF filament that brings the attributes of this industrial plastic to 3D printing applications.

8.3.1.1.10 Polyetherimide (PEI) PEI stands for polyetherimide, a polymer that was first developed by Joseph Wirth of the Plastic Division of GE back in the early 1980s. It is an amorphous plastic with a characteristic semi-transparent amber color. PEI has the advantages of high-temperature resistance, good dimensional stability, as well as good chemical resistance, good flame retardancy, strong electrical properties, high strength, and high rigidity. Adding glass fiber, carbon fiber, or other fillers can achieve the purpose of reinforcement modification, but also with other engineering plastics to form a heat-resistant polymer alloy, is widely used in aviation and electrical and electronic industries.

PEI has a glass transition temperature of 217 °C and starts to melt at 240 °C. This makes PEI an excellent material for functional parts that can endure a lot of friction and medical tools that need to be autoclaved regularly. The chemical stability of PEI also allows its use in versatile applications, particularly for parts that are often used outdoors and exposed to UV radiation. PEI resists attacks by acids, bases, and oils, making it a valuable material for various industrial and commercial settings.

8.3.1.2 Semi-crystalline Polymers

8.3.1.2.1 Polypropylene (PP) PP is a homopolymer member of polyolefins and one of the most widely used low-density and low-cost thermoplastic semi-crystalline polymers. The applications are generally used in different industries such as the military, household appliances, cars, and construction because of their physical and chemical properties. However, PP has low thermal, electrical, and mechanical properties compared to other engineering plastics (PC and PA) and has a high coefficient of friction in dry shear conditions. The mechanical properties of PP are improved by combining with inorganic fillers in the form of nanoparticles. PP exhibits great chemical resistance to dilute or concentrated acids, alcohols, bases, and mineral oils, and thus its printed components are appropriate candidates for corrosion-resistant applications. In addition, PP is biocompatible, and its components could be used as permanent surgical implants.

8.3.1.2.2 Polyethylene (PE) PE is one of the most popular thermoplastic materials. This is a lightweight, durable plastic that is often used for frozen food bags, bottles, cereal liners, and yogurt containers. It is available in different crystalline structures, referred to as HDPE, LDPE, and LLDPE.

The polymers are physiologically safe, tasteless, and odorless. PE is resistant to most acids and alkalis, but also to oils, fats, petrol, and aliphatic hydrocarbons. This commodity plastic is produced by addition or radical polymerization. Accordingly, exposure to sunlight will lead to embrittlement of the PE-printed parts. PE is a good electrical insulator, but it is very easy to charge electrostatically. PE has low strength, rigidity, and hardness, but has high ductility and impact strength and low sliding friction.

8.3.1.2.3 Polycaprolactone (PCL) PCL is one of the polymers that have been attempted for use in 3D printing applications. It has an extremely low print temperature and is suitable for 3D doodling. This means PCL can easily be 3D printed on most filament-based 3D printers with low hot-end temperatures. PCL has an excellent attachment which results in enhanced improvements in toughness, durability, and strength. It has a smooth and matt surface. PCL softens in the water of more than 50 °C and can be reshaped. It hardens slowly when cooling down to room temperature. PCL can be recycled. PCL has properties different than other commonly used 3D printing plastics, some of which make it unique and applicable for unusual applications. Due to its excellent biocompatibility and biodegradability, PCL is most extensively used as a FFF feedstock polymer for biomedical applications, mainly for the production of scaffolds or tissue engineering.

8.3.1.2.4 Polybutylene Terephthalate (PET, PETE) Polyethylene terephthalate (PET, PETE) is a polyester-based material that combines excellent mechanical, electrical, and thermal properties with very good chemical resistance and dimensional stability. PET also has low moisture absorption features and good flow properties, making it a great material to use for waterproof containers such as food and beverage storage. It is a great material to use for making lightweight products such as potted planters, insulated bottles, or Tupperware. Recycled PET might be converted into sheets for packaging, cloths, fibers, and automotive pieces.

8.3.1.2.5 Thermoplastic Polyurethane (TPU) Thermoplastic polyurethane (TPU) is a thermoplastic flexible material with rubber-like properties, developed for the production of durable elastomer parts. The material exhibits high elongation, superior toughness, durability, and abrasion resistance. It enables prototyping of high-functioning, durable, and complex parts with the expected material properties found in an elastomeric material, including superior tear resistance, fatigue, and memory recovery. TPU brings the benefits of elastomers to FFF 3D printing and offers the capability to quickly produce large and complex elastomer parts. Typical applications include flexible hoses, tubes, air ducts, seals, protective covers, and vibration dampeners.

8.3.1.2.6 Polylactic Acid (PLA): Polylactic acid (PLA) is a polymer obtained by polymerizing lactic acid as the main raw material. It can be just a biodegradable, biocompatible, and compostable polyester based on renewable resources like corn, cane molasses, and beet sugar levels. The production process of PLA is pollution-free, and the product is biodegradable and can be recycled in nature, making it an ideal green polymer material. Polylactic acid has good thermal stability with processing temperatures of 170–230 °C. It has good solvent resistance and can also be processed in a variety of ways, such as extrusion, spinning, biaxial stretching, and injection blow molding. Products made from PLA are biodegradable, in addition to good biocompatibility, gloss, transparency, feel, and heat resistance. Some modified PLA also has certain antibacterial, flame retardant, and UV resistance properties. Therefore, it is widely used, mainly in construction, agriculture, and medical or health fields.

8.3.1.2.7 Polyamide (Nylons, PA) Polyamide (PA), commonly known as nylon, is a generic term for polymers containing acyl groups in the repeating units of the macromolecular main chain. Polyamide can be produced by ring-opening polymerization of endo-acid gum, or by condensation of diamines with diacids. Polyamides were first developed by DuPont in the United States for fiber resins and industrialized in 1939. In the 1950s, it began to produce injection-molded products to replace metal and meet the requirements of downstream industrial products for lightweight and cost reduction.

PA has comprehensive properties, including mechanical properties, heat resistance, wear resistance, chemical resistance, and self-lubrication, and it also has a low friction coefficient, certain flame retardancy, easy processing. It can be filled with other fillers for enhancement and modification to improve performance and expand the scope of the application. There are many varieties of PA, including PA6, PA66, PAII, PA12, PA46, PA610, PA612, and PAI010 as well as new varieties such as semi-aromatic nylon (PA6T) and special nylon developed in recent years.

8.3.1.2.8 Polyoxymethylene (POM) Polyoxymethylene (POM) is an engineering thermoset plastic. It is a nylon polymer manufactured and used since the early 1920s. POM filament is among the widely utilized materials after polyamide (PA) and PP. Its color is white. It has been used for centuries in the textile industry, but it is also gaining traction in 3D prints. POM can be used to create parts with high mechanical properties, which remain stable and not deformed even after getting exposed/disclosed to temperatures ranging from −40 to +140 °C. It is very resistant to moisture, making POM a far superior choice in wet and high-humidity environments. It also has excellent electrical insulation. Thanks to its impact, abrasion resistance, low friction, and excellent sliding properties, it is ideal for use in places that require a lot of movement. It is an essential component in manufacturing smaller gear wheels with features such as backpack buckles. Heat-resistant makes POM suitable for appliances subjected to heat, like heating stoves. It is also used to produce components, such as automobile inside structure fittings or implants. It is compelling in medicine because of its chemical and mechanical properties, including resistance to disinfectants, durability, and color.

8.3.1.2.9 Polyphenylene Sulfide (PPS) PPS is a high-performance polymer. Due to its structure, PPS exhibits exceptional chemical resistance along with high thermal and mechanical properties. In addition to its low water absorption, it also has good dimensional stability and excellent electrical properties. PPS can fabricate custom bottles, packaging, and other parts for chemical processing. Made of PPS, these filaments will not break down even after prolonged exposure to virtually any acid, base, or solvent. They also stand up to oil and fuel, so they are sometimes made into automotive parts. Because they have a low density and will not transfer electricity, they make parts that are lightweight and protect components from electric currents.

8.3.1.2.10 Polyetheretherketone (PEEK) Polyetheretherketone (PEEK) is a polymer consisting of a ketone bond and two ether bonds in the main chain structure of the repetitive unit. PEEK has high-temperature resistance and chemical corrosion resistance, with a melting point of 334 °C, softening point of 168 °C, and tensile strength of 132–148 MPa. It can be used as high-temperature resistant structural materials and electrical insulation materials. Combined with glass fiber or carbon fiber, the PEEK polymers can also prepare reinforcing materials. Due to the excellent comprehensive properties of PEEK, traditional materials such as metals and ceramics can be replaced in many special applications. The resin's high-temperature resistance, self-lubrication, wear resistance, and fatigue resistance make it one of today's most popular high-performance engineering plastics, mainly used in aerospace, automotive industry, electrical electronic, and medical devices.

8.3.2 Classification of Polymer Composites

Since the mechanical properties of pure thermoplastic materials are not suitable for some applications, it is necessary to change the mechanical properties of pure thermoplastic parts. The addition of fiber fillers or nanoparticles to a polymer matrix can significantly improve the properties of 3D-printed parts and obtain the desired mechanical or functional properties. The polymer composites developed so far can be divided into structural polymer materials and functional polymer materials according to the purpose of use.

In structural and functional polymer matrix composites, the materials are mainly composed of two parts: reinforced phase and matrix material [5]. Generally, the continuous phase in the composite structure is called matrix, which mainly plays the role of supporting the strong/protective phase and transmitting loads between the reinforced phases, such as ABS, PLA, PA, PP, PE, TPU, PPI, PPS, and PEEK. The dispersed phase distributed in the matrix in an independent form is called a reinforced phase due to its characteristics of significant reinforcing material properties, such as fibers and conductive particles.

8.3.2.1 Structural Polymer Matrix Composites

Structural polymer matrix composites are mainly used in the manufacture of force-bearing components. According to the length of the fibers, the composites can be classified as short fiber-reinforced composites, long fiber-reinforced composites, and continuous fiber-reinforced composites.

8.3.2.1.1 Short Fiber Reinforced Composites The short fibers are uniformly and irregularly distributed in the composites with high formability after FFF. In terms of performance, after adding short fiber reinforcement, the performance of the composite material will change greatly, and there will usually be a great improvement in mechanical properties. At the same time, some special staple fibers can also be added to improve the functionality of composite materials, such as PA 6 with glass fiber reinforced, its heat deformation temperature can be increased from 50 to 190 °C. The electrical conductivity of the composite material can also be improved after being reinforced with carbon fiber.

8.3.2.1.2 Long Fiber Reinforced Composites Long fiber reinforced composites have made great progress in mechanical properties, chemical properties, electrical properties, and physical properties in the last decade. The researchers have conducted extensive research on key technologies such as the composite process of long fiber reinforced composites, the mechanical properties of composites, and the influence of fiber and resin interfaces (such as adhesion and fiber orientation) on the properties of composite materials. Compared with the printed parts of the short fibers reinforced composites, long fiber-reinforced composites have better mechanical properties due to their better fiber length and fiber dispersion such as specific stiffness, specific strength, impact resistance, creep resistance, and fatigue resistance of the composites.

8.3.2.1.3 Continuous Fiber Reinforced Composites Unlike short and long-fiber-reinforced composites, the reinforcing fibers of continuous fiber-reinforced composites are continuous, and their mechanical properties are much higher than those of long-fiber-reinforced composite parts. Due to the improvement of pre-impregnation technology, composites with higher fiber content can be prepared, and their reinforcement effect can then be improved more significantly, and even the mechanical properties of structural composite materials in the aerospace field have been fully met. Compared with the former two fiber-reinforced composites, the forming of continuous fiber-reinforced composites is more complex, and traditional injection molding, extrusion molding, and other processes cannot achieve the manufacture of continuous fiber-reinforced composite components. Therefore, emerging manufacturing technologies such as AM processes and fiber placement for continuous fiber-reinforced composites have paid more and more attention in recent years.

8.3.2.2 Functional Polymer Matrix Composites

Functional polymer matrix composites provide other physical or chemical properties in addition to mechanical properties. In general, the reinforcing phase of functional composites can be realized by adding particles or filler materials with optical, electrical, magnetic, thermal, chemical, and biochemical. The main functional body of functional composites can be composed of one or more functional materials. The composite material with multiple functional bodies can have multiple functions. Composites with multifunctional bodies can have multiple functions.

8.4 Physicochemical Characterization of 3D-printed Parts

Various thermoplastics are used as feedstocks for FFF technology. As physical and chemical properties of thermoplastics are different, it is essential to know the physicochemical characterization of 3D-printed parts' properties to choose an appropriate material according to its function and purpose.

8.4.1 Physical Properties of 3D-printed Parts

Physical properties of polymers include mechanical properties, thermal properties, electrical, magnetic, and optical properties. Some of these are discussed herewith in the following sections.

8.4.1.1 Mechanical Properties

The ability of a polymer to resist or withstand physical stress is known as its mechanical property. In general, the mechanical properties of FFF 3D prints refer to the macroscopic properties of standard mechanical test pieces printed in accordance with ISO or ASTM international standards, resisting deformation, and fracture under the action of external forces.

Understanding the mechanical properties of different applicable filament materials is one of the critical factors in deciding where a particular kind of polymer should be used to fabricate the 3D printed parts in FFF. The main reference indicators are stiffness, plasticity, toughness, strength, hardness, fatigue, and strength. These mechanical properties need to be determined on the material standard testing machine according to the test methods and procedures using standard specimens, and the stress-strain curve of each material can be determined at the same time.

8.4.1.1.1 Material Yield, Fracture, and Strength Properties The mechanical behavior of a printed standard mechanical test parts, which develops from test deformation to yielding, and even fracture under the continuous action of a larger external force or the short duration of a strong external force, will involve mechanical strength issues. Some of the parameters are discussed herewith in the following sections.

Strength: The strength of FFF printed polymer materials to elongate without breaking is its tensile or bending strength. Various types of strength of polymers are tensile strength, flexural strength, compressibility, and flexibility. The increasing orders of different polymers' strength are linear < branched < cross-linked < network. Factors that affect its strength are:

a. Molecular Weight: The tensile strength of the polymer arises with an increase in molecular weight and reaches a level of saturation after a specific value.
b. Cross-linking: The motion of the chains in polymers is restricted by cross-linking and increases the strength of the polymer.
c. Crystallinity: In the crystalline phase, the intermolecular bonding is stronger and more significant. Hence, the crystallinity of the polymer increases its strength.

Elongation at Break: Percentage elongation of FFF printed polymer parts to break is the measure of elongation of a polymer without deforming it.

Young's Modulus: It is the ratio of tensile stress and tensile strain. It determines how easily test specimens of polymers can stretch and deform. In short, it is the measure of the stiffness of a polymer.

Toughness: The toughness of a polymer is determined by the area under the stress-strain curve. The toughness measures the energy absorbed by the material before it breaks.

For polymer materials used in FFF printing, the strength and modulus of polymer materials vary greatly from commodity polymers to high-performance thermoplastic polymers, as shown in Table 8.1. For example, for amorphous commercial ABS polymer filament, its tensile and flexural strength is only 28.8 and 65.5 MPa, respectively, its figure being about 50% lower than the amorphous high-performance polymer of ULTEMTM 1010. For the semi-crystalline polymers, Young's modulus of PEEK is 18-fold higher than that of PP. ABS-ESD7™ (acrylonitrile butadiene styrene-electrostatic dissipative) is an ABS thermoplastic, which adds the electrostatic dissipative particles with static dissipative properties suited for static discharge-sensitive applications. The material has similar mechanical properties to ABS pure polymers. ABS-ESD7 prevents static electricity

Table 8.1 Mechanical properties of typical polymers and polymers composite filament.

Material	Strength (MPa)		Young's modulus (MPa)	Elongation at break (%)	Company
	Tensile	Flexural			
ABS	28.80	65.50	1847	3.8	INTAMSYS
PPSU/PPSF	55.00	110.00	2100	3.0	Fortus
ULTEMTM1010	79.20	128.00	3040	4.0	Stratasy
PP	8.70	13.00	220	18.0	Ultimaker
PCL	45.00	18.00	350	15.0	Facilan
PLA	45.60	87.70	2641	2.4	INTAMSYS
PA	66.20	97.00	2223	9.9	IEMAI 3D
PEEK	98.00	125.00	4000	6.2	VICTREX
ABS-ESD7	31.80	60.40	2120	2.4	Stratasys
PA6-CF	105.0	169.00	7453	3.0	PolyMide

Note: All data is as of August 2022 and sourced from respective company websites.

buildup so it will not produce a discharge or attract other materials like powders, dust, and fine particles. The material is ideal for jigs and fixtures used to fabricate and assemble electronic components and associated production line and conveyor parts. It is also useful for producing functional prototypes, enclosures, and packaging.

8.4.1.1.2 Hardness Hard polymers resist the penetration of hard substances into them. They withstand wear and tear, and scratches, and are used in the manufacturing of constructing devices.

8.4.1.1.3 Viscoelasticity Polymers generally exhibit viscoelastic behavior, which is a combination of viscous and elastic behavior. Viscoelastic behavior is of paramount importance in defining the efficiency of the polymer printing process. Most polymers are thixotropic and their viscosity decreases with increasing process temperature. For smooth extrusion, a process temperature should be selected that ensures low viscosity of the formulation blend.

Formulants with high melt viscosity cannot be extruded through a 3D printer nozzle because the molten material has poor flow. Thus, appropriate formulation blends are required to reduce melt viscosity and permit extrusion to be performed at lower operating temperatures. The addition of plasticizers reduces the viscosity and improves molten mass extrudability. An increase in 3D printer nozzle temperature may reduce viscosity but could also affect material degradation [6].

8.4.1.1.4 Rheological Characteristics Rheology is a new branch of mechanics, which mainly focuses on the deformation and flow of materials under the conditions of stress, strain, temperature, humidity, radiation, and other factors related to time factors. Thus, studying rheological properties of a polymer provides enough information on FFF process. The rheological behavior of polymers is characterized by experimental studies using various rheometers, such as rotational rheometers, dynamic shear rheometers, drop plates, hybrid rheometers, and capillary rheometers, which have been widely used for the flow characterization of polymeric materials [7].

The main factors affecting the rheological properties of the polymers are shape, size, nature, interfaces, orientation, and viscosity of the particles. The dispersion size of the fillers is larger, which reduces the matrix viscosity. Proper dispersion and interface between the materials and matrix produces high-quality material with desired properties [8]. For BF/PEEK composites, suitable fiber content parameters for 3D printing can be studied by studying their viscoelastic behavior such as energy storage modulus (G'), plural viscosity ($|\eta^*|$), and loss factor ($\tan\delta$) [9].

8.4.1.2 Thermal Properties

Thermal properties play a large role in the behavior of polymers, causing them to react in different ways. The temperature of a polymer can greatly affect the way it responds to stress. In the amorphous region of the polymer, at lower temperatures, the molecules of the polymer are in, say, a frozen state, where the molecules can vibrate slightly but are not able to move significantly. This state is referred to as the glassy state. In this state, the polymer is brittle, hard, and rigid, analogous to glass. Hence the name glassy state. The glassy state is similar to a supercooled liquid where the molecular motion is in the frozen state. The glassy state shows hard, rigid, and brittle nature analogous to a crystalline solid with molecular disorder as a liquid. If the polymer is heated, there will be a lot of kinetic energy, and the polymer chains can wiggle around each other, and the polymer becomes soft and flexible, similar to rubber. This state is called the rubbery state. The temperature at which the glassy state makes a transition to rubbery state is called the glass transition temperature (T_g). Note that the glass transition occurs only in the amorphous region, and the crystalline region remains unaffected during the glass transition in the semi-crystalline polymer.

8.4.1.2.1 Crystallinity The polymeric molecular chains found in the polymer are mainly in two forms which are as follows: crystalline form and amorphous form. The chains in crystalline form are folded and make a lamellar structure arranged in the regular manner, and in amorphous form, the chains are in irregular manner. The lamellae are embedded in the amorphous part and can communicate with other lamellae via tie molecules. Polymers may be amorphous or semi-crystalline in nature.

$$\text{The crystallinity is given by: Crystallinity} = \frac{\rho_c(\rho_s - \rho_a)}{\rho_s(\rho_c - \rho_a)} \times 100\%$$

ρ_c – density of the completely crystalline polymer,
ρ_a – the density of the completely amorphous polymer,
ρ_s – density of the sample.

A typical range of crystallinity can be defined as amorphous to highly crystalline. The polymers having simple structural chains, as linear chains and slow cooling rate will result in good crystallinity as expected. In slow cooling, sufficient time is available for crystallization to take place. Polymers having high degree of crystallinity are rigid and have high melting points, but their impact resistance is low. However, amorphous polymers are soft and have lower melting points. For a solvent, it is important to state that it can penetrate the amorphous part more easily than the crystalline part.

8.4.1.2.2 Thermal Expansion Thermal expansion is the concept of materials taking up more space as the temperature increases. The reason for this is that as the temperature increases, the kinetic energy increases, and the atoms vibrate over a longer distance. The extent to which a polymer expands or contracts when subjected to heat or cold is measured by this property. The thermal expansion can be measured using thermomechanical analysis. Thermal expansion is important in FFF to understand the tolerances of 3D printed parts. The polymer passing through the extruder die impacts filament suitability. High thermal expansion of the filament increases its diameter and may hinder its processing in a 3D printer. Extrudate contraction reduces filament diameter and results in brittle filament that may be unfit for printing.

8.4.1.2.3 Melting Point and Glass Transition Temperature The glass transition temperature of polymer, melting point, and degradation temperature are crucial for FFF 3D printing filament with consistently high quality. Glass transition temperature, melting point, and material degradation are evaluated by differential scanning calorimetry (DSC) and thermogravimetric analysis (TGA). The DSC curve can be seen in Figure 8.3.

The glass transition temperature is the property of the amorphous region of the polymer, whereas the crystalline region is characterized by the melting point. The value of glass transition temperature is not unique because the glassy state is not in equilibrium. The value of glass transition temperature depends on several factors such as molecular weight, measurement method, and the rate of heating or cooling.

The semi-crystalline polymer shows both the transitions corresponding to their crystalline and amorphous regions. Thus, the semi-crystalline

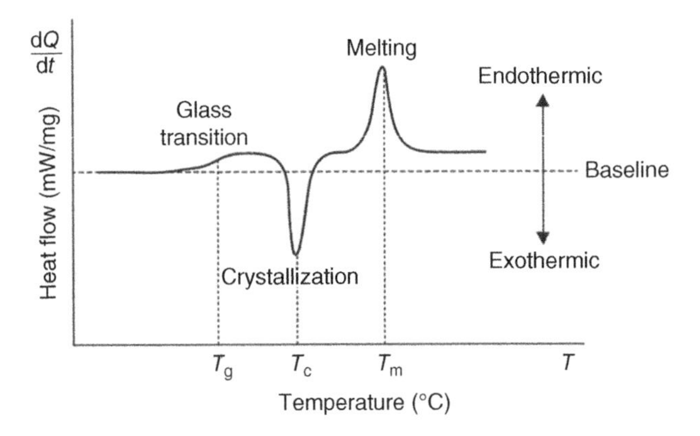

Figure 8.3 Typical DSC curve for a polymer.

polymers have true melting temperatures (T_m) at which the ordered phase turns to disordered phase, whereas the amorphous regions soften over a temperature range known as the glass transition (T_g). It should be noted that amorphous polymers do not possess the melting point, but all polymers possess the glass transition temperature. The polymer melting point is increased if double bonds, aromatic groups, bulky or large side groups are present in the polymer chain because they restrict the flexibility of the chain. The branching of chains causes the reduction of melting point, as defects are produced because of the branching.

Crystalline polymers are characterized by a melting point T_m and amorphous polymers are characterized by a glass transition temperature T_g. For crystalline polymers, the relationship between T_m and T_g has been described by Boyer as follows:

$$\begin{cases} \dfrac{T_g}{T_m} \approx \dfrac{1}{2}, & \text{symmetrical polymers} \\ \dfrac{T_g}{T_m} \approx \dfrac{2}{3}, & \text{unsymmetrical polymers} \end{cases}$$

It was found that many polymers with a T_g/T_m ratio below 1/2 are highly symmetrical and consist of small repeating units of one or two main-chain atoms, each carrying only single-atom substituents, such as PE and POM polymer filament. These polymers are markedly crystalline. Polymers with T_g/T_m ratios above 2/3 are usually unsymmetrical. They can be highly crystalline if they have long sequences of methylene groups or are highly stereo-regular. The majority of the polymers, however, have T_g/T_m ratios between 0.5 and 0.75 with a maximum number around 2/3; both symmetrical and unsymmetrical polymers belong to this group.

8.4.1.2.4 Thermal Conductivity Thermal conductivity is the rate of heat transfer by conduction. This decides the extent to which a polymer acts as an insulator of heat. The stiffness of molecules decides whether a polymer is a good conductor of heat. Thermal conductivity mainly depends on the orientation of chain segments, the density of polymer, temperature, structure, and degree of crystallinity of 3D printing polymer filaments. For example, the cell size of foamed polymer may affect thermal conductivity. Smaller foam cell size tends to lower thermal conductivity. Most foamed polymers have thermal conductivity values in the order of, which is about 10 times less than the same polymers. A polymer filament with a highly crystalline and ordered structure will have higher conductivity than an amorphous polymer.

8.4.1.3 Electrical and Optical Properties

In addition to the mechanical and thermal properties of polymers, some polymer filaments also have dielectric properties, electrostatic, piezoelectric, and pyroelectric properties breakdown behavior, and ionic conductivity. Polymer parts of FFF have been widely used in cable insulation, film capacitors, circuits, electronic circuits, switching devices, electrostatic copying, energy storage and conversion, and other aspects.

8.4.1.3.1 Arc Resistance Arc resistance is the ability of the polymer filament to resist the action of a high-voltage electrical arc and resist the formation of a conducting path along its surface under a given time. It is used for differentiation among similar materials due to their resistance to the action of high voltage low current to the surface of the insulation. It is also a critical property when selecting polymer for insulation applications since the loss of insulation can lead to devastating consequences in certain application areas.

8.4.1.3.2 Refractive Index The refractive index is an important optical property of polymers and is widely used in material science. Polymers use this property in spectroscopy. The knowledge of the refractive index is crucial in all optical applications of transparent polymers. Since it is characteristic of each material, it can be used for identification purposes or the prediction of other properties. For example, the refractive index undergoes a second-order transition at the glass transition temperature and thus can be used to determine its value. The refractive index is directly related to polarizability and depends on the wavelength of light.

8.4.2 Chemical Properties

The chemical properties of FFF filament polymers are dependent on their chemical and physical structures and molecular weight.

8.4.2.1 Molecular Weight

Polymers are composed of many constituent monomers that are chemically linked together. The process of polymerization results in polymer chains of varying lengths; this gives rise to polymer chains with different molecular weights and molecular weight distributions. The molecular weight of a polymer is described as an average molecular weight due to the different lengths of polymer chains [10]. Mechanical (stress, strain, impact, fatigue, and creep) and thermal properties are influenced mainly by molecular weight and molecular weight distribution. Electrical properties such as conductivity, dielectric constant, and dielectric loss as well as other physical properties also depend on the molecular weight significantly [11], including osmotic pressure, viscosity, depression in freezing point, and elevation in boiling point.

8.4.2.2 Chemical Permeability

Chemical permeability is the tendency of particles to pass through the polymers. The size of the gas or liquid molecule is extremely important. Small molecules can work their way through the polymer much more easily than large molecules. Size effect, therefore, can outweigh all other permeation effects. The polymer with many polar groups is sensitive to a polar chemical, that same polymer would be permeable to a polar gas or liquid. Conversely, a nonpolar polymer would be a barrier to polar gases and liquids. For example, PP has very low water permeation but relatively high oxygen permeation. PVA polymer filament has very low oxygen permeation but relatively high water permeation. Low-density PE is less permeable to air, so it is used to pack food items. The amorphous polymer exhibits higher gas and liquid permeation than that of the crystalline polymer due to the higher density of crystalline.

8.4.2.3 Chemical Resistance

Chemical resistance is the ability of a polymeric 3D printing filament to maintain its original properties after being exposed to a chemically aggressive environment for a specified period. A wide spectrum of polymers exists that can resist acids, solvents, alcohol, hot water, and other substances, which are most suitable for use under demanding circumstances. Compared with metal, most polymers are corrosion-resistant because it does not react with water. They are usually used as a protective coating for metal. Polyester has good film formation characteristics and is a good coating material. In general, engineering materials are the most resistant to chemicals. Those with excellent chemical resistance include PP, PA, PE, and PEEK.

8.4.2.4 Chemical Degradability

Polymers can be degraded by the action of decomposition, microorganisms, or sunlight. Natural polymers like rubber are biodegradable, while synthetic polymers are non-biodegradable. Polymers can be made to degrade photochemically by incorporation of carbonyl groups that absorb ultraviolet (UV) radiation to form excited states energetic enough to undergo bond cleavage. Microorganisms degrade polymers by catalyzing hydrolysis and oxidation. By the combination of sunlight and microorganisms, the degradation of polymers will be more effective. Many biodegradable polymers are used for sutures, drug delivery, and tissue engineering.

8.5 Effect of Phase Change on the Quality of 3D-Printed Parts

8.5.1 The Factors that Affect the Crystallization of 3D-Printed Parts

Crystallinity means the degree of regular and structural order in a material or a molecular structure. Crystalline materials have long-range order in their structure. Crystallinity is a common polymer morphology. When the crystallinity of a structure is low, we can say that it is more amorphous. Three main factors affect the crystallinity of the polymer.

First, polymer structure can influence the crystallization of 3D-printed parts. However, the molecular varies from the filament to filament of 3D printing polymers. If the polymer molecule is regular and in order, it will pack into a crystal easily.

Second, intermolecular force is one of the factors affecting crystallinity. The presence of polar side groups helps to form strong intermolecular interactions. So, polymer chains can come closer and pack tightly. For example, PA can form strong hydrogen bonds between the nitrogen-bonded hydrogen of one nylon chain and the carbonyl oxygen of another nylon chain. These hydrogen bonds help to pack nylon chains closely. As a result, the crystallinity of nylon increases.

Third, external effects (3D printing processing conditions) affect the crystallinity of 3D printed parts. The crystallinity of polymer products varies due to processing conditions, such as the cooling rate of melting material and environment temperature. When the cooling rate is high, polymer molecules do not have time to organize in orderly, thus the crystallinity of the printed part is relatively low. Also, the polymer molecule of printed has sufficient time to arrange orderly if the ambient temperature (chamber, platform, and nozzle) is high, and that results in a higher degree of crystallinity.

8.5.2 The Effect of Crystallinity on Physical Properties

The degree of crystallinity has great influence on the physical properties of polymer materials. Many physical properties such as mechanical properties, optical properties, thermal properties, and electronic properties of polymers are related to their degree of crystallinity.

8.5.2.1 Optical Properties

The degree of crystallinity affects the optical absorption properties of polymers. The chains of polymers are packed tightly when the printed parts with high crystallinity. Consequently, light rays cannot go through these highly crystalline regions easily. Most of the light rays scatter on a highly crystalline polymer surface. But light rays can easily transmit through the more amorphous polymer. The crystallinity of the composite polymer increases, leading to increased optical absorption in the low-photon-energy region [12]. More crystalline polymers are opaque or translucent. With the increment of amorphous content, they become more transparent. For example, 3D printed parts made of PEEK polymer materials are opaque, light brown in color for high crystallinity, and translucent amber for low crystallinity, as shown in Figure 8.4.

8.5.2.2 Thermal Properties

The crystallinity of a polymer affects its thermal properties. For instance, when the degree of crystallinity is increased in a polymer, the intensity of the glass transition is decreased together with a broader transition temperature range [13]. In FFF process, as the temperature increases, polymer filament of hard and rigid solid gradually obtains sufficient thermal energy to enable its chains to move freely and behave like a viscous liquid. According to the crystallinity of polymer, there are two ways that polymer can pass from solid to the liquid state. As the temperature increases, amorphous polymers pass through the glass transition (T_g) where glassy-state polymer becomes a rubbery state. When the temperature continues to increase, the rubbery polymer turns into a viscous liquid. In the case of semi-crystalline polymers, it has both glass transition point and melting point due to their crystalline and amorphous properties. In addition, the increasing crystallinity makes an improvement in heat resistance of polymer specimens.

8.5.2.3 Water Absorption and Wear Resistance

Consistent with the principle of polymers' optical absorption properties, water molecules cannot penetrate through polymer chains of high crystallinity easily. The water absorption rate of polymers in highly crystalline

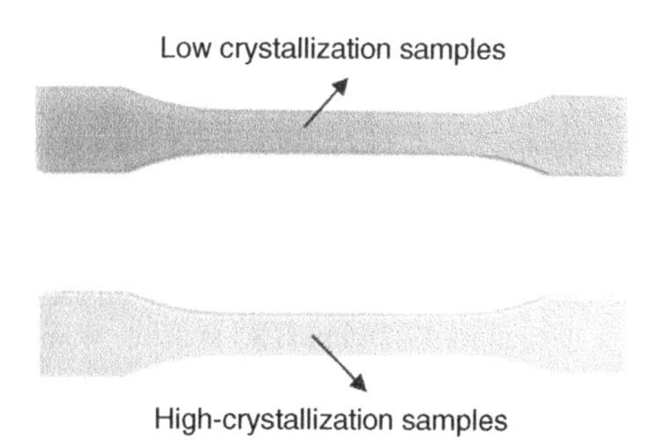

Figure 8.4 Different crystallinity of 3D-printed PEEK samples.

is small. As the crystallinity decreases, so does the water absorption of the polymer [14]. The water vapor permeability of unoriented region has been shown to increase as the amorphous fraction is increased [15].

The wear resistance of polymers was enhanced with the increase of crystallinity in rubbery state since larger energy was needed to break the links between crystallites [16]. Scholars have studied the links between tribological behavior and degree of crystallinity of PTFE composites [17].

8.5.2.4 Mechanical Properties

The crystallinity of polymers contributes significantly to the mechanical properties of the polymer [18–20], including strength, stiffness, hardness, and toughness.

Increasing the crystallinity of a polymer increases the strength, stiffness, yield point, tensile strength, and hardness [21], but reduces the impact strength, particularly in the high crystallinity range [22].

The crystalline phase has more intermolecular bonds, and without affecting the fracture toughness of the material, the high crystallinity of the 3D-printed part will increase its tensile strength, yield strength, flexural strength, tensile modulus, and flexural modulus [23]. However, high crystallinity limits the molecular movement of polymer chains. Polymers with higher crystallinity have high-order polymer chains that are tightly packed, so the resistance of local deformation (hardness) of polymers is very high with greater brittleness. At this time, the tensile fracture strength and impact resistance of 3D printing test pieces are reduced. Thus, toughness decreases as the crystallinity of the polymer increases.

References

1 Xiong, Y., Tang, Y., Zhou, Q. et al. (2022). Intelligent additive manufacturing and design state of the art and future perspectives. *Additive Manufacturing* 59: 103139.

2 Gibson, I., Rosen, D.W., Stucker, B. et al. (2021). *Additive Manufacturing Technologies*. Springer.

3 Xu, X., Ren, H., Chen, S. et al. (2023). Review on melt flow simulations for thermoplastics and their fiber reinforced composites in fused deposition modeling. *Journal of Manufacturing Processes* 92: 272–286.

4 Tlegenov, Y., Hong, G.S., and Lu, W.F. (2018). Nozzle condition monitoring in 3D printing. *Robotics and Computer-Integrated Manufacturing* 54: 45–55.

5 Schouten, M., Wolterink, G., Dijkshoorn, A. et al. (2020). A review of extrusion-based 3d printing for the fabrication of electro-and biomechanical sensors. *IEEE Sensors Journal* 21 (11): 12900–12912.

6 Bandari, S., Nyavanandi, D., Dumpa, N., and Repka, M.A. (2021). Coupling hot melt extrusion and fused deposition modeling: critical properties for successful performance. *Advanced Drug Delivery Reviews* 172: 52–63.

7 Prabhu, R. and Devaraju, A. (2021). Recent review of tribology, rheology of biodegradable and FDM compatible polymers. *Materials Today Proceedings* 39: 781–788.

8 Rueda, M.M., Auscher, M.-C., Fulchiron, R. et al. (2017). Rheology and applications of highly filled polymers: a review of current understanding. *Progress in Polymer Science* 66: 22–53.

9 Wang, B., Yu, S., Mao, J. et al. (2021). Effect of basalt fiber on tribological and mechanical properties of polyether-ether-ketone (PEEK) composites. *Composite Structures* 266: 113847.

10 Umoren, S.A. and Solomon, M.M. (2016). Polymer characterization: polymer molecular weight determination. *Polymer Science: Research Advances, Practical Applications and Educational Aspects, Formatex Research Center SL* 1: 412–419.

11 Singh, A. (2017). A review of methods of molecular weight determination of polymers. *AKGEC International Journal of Technology* 8 (2): 61Ā4.

12 Erb, T., Zhokhavets, U., Gobsch, G. et al. (2005). Correlation between structural and optical properties of composite polymer/fullerene films for organic solar cells. *Advanced Functional Materials* 15 (7): 1193–1196.

13 Herrera, N., Mathew, A.P., and Oksman, K. (2015). Plasticized polylactic acid/cellulose nanocomposites prepared using melt-extrusion and liquid feeding: mechanical, thermal and optical properties. *Composites Science and Technology* 106: 149–155.

14 Tsuji, H., Okino, R., Daimon, H., and Fujie, K. (2006). Water vapor permeability of poly (lactide) s: effects of molecular characteristics and crystallinity. *Journal of Applied Polymer Science* 99 (5): 2245–2252.

15 Lasoski, S. Jr. and Cobbs, W. Jr., (1959). Moisture permeability of polymers. I. Role of crystallinity and orientation. *Journal of Polymer Science* 36 (130): 21–33.

16 Song, F., Wang, Q., and Wang, T. (2016). The effects of crystallinity on the mechanical properties and the limiting PV (pressure×velocity) value of PTFE. *Tribology International* 93: 1–10.

17 Conte, M., Pinedo, B., and Igartua, A. (2013). Role of crystallinity on wear behavior of PTFE composites. *Wear* 307 (1): 81–86.

18 Marshall, R. (1994). Effect of crystallinity on PVC physical properties. *Journal of Vinyl Technology* 16 (1): 35–38.

19 Xu, Q., Xu, W., Yang, Y. et al. (2022). Enhanced interlayer strength in 3D printed poly (ether ether ketone) parts. *Additive Manufacturing* 55: 102852.

20 Dusunceli, N. and Colak, O.U. (2008). Modelling effects of degree of crystallinity on mechanical behavior of semicrystalline polymers. *International Journal of Plasticity* 24 (7): 1224–1242.

21 Chivers, R.A. and Moore, D.R. (1994). The effect of molecular weight and crystallinity on the mechanical properties of injection moulded poly(aryl-ether-ether-ketone) resin. *Polymer* 35 (1): 110–116.

22 Starkweather, H.W. Jr.,, Moore, G.E., Hansen, J.E. et al. (1956). Effect of crystallinity on the properties of nylons. *Journal of Polymer Science* 21 (98): 189–204.

23 Yang, C., Tian, X., Li, D. et al. (2017). Influence of thermal processing conditions in 3D printing on the crystallinity and mechanical properties of PEEK material. *Journal of Materials Processing Technology* 248: 1–7.

9

3D Printing Optimization: Importance of Rheological Evaluation in 3D Printing

Abbas Tcharkhtchi[1], Reza Eslami Farsani[2], and Hamid Reza Vanaei[3,4,5]

[1]*Arts et Métiers Institute of Technology, CNRS, CNAM, PIMM, HESAM University, 75013, Paris, France*
[2]*Faculty of Mechanical Engineering, K. N. Toosi University of Technology, Tehran, Iran*
[3]*ESILV, Léonard de Vinci Pôle Universitaire, 92916 Paris La Défense, France*
[4]*Léonard de Vinci Pôle Universitaire, Research Center, 92916 Paris La Défense, France*
[5]*Arts et Métiers Institute of Technology, CNAM, LIFSE, HESAM University, 75013 Paris, France*

9.1 Introduction

Fused filament fabrication (FFF), which is one of the additive manufacturing (AM) techniques, involves creating parts by depositing molten filament of a thermoplastic polymer, resulting from extrusion, onto a plate with precision. This process is controlled by a computer according to a previously recorded program. FFF includes the deposition of polymer filaments in successive x–y planes along the z direction, creating a layer-by-layer deposition mechanism. Among the wide range of available thermoplastics, only a limited number of polymers, such as polylactic acid (PLA) and acrylonitrile butadiene styrene (ABS), as well as polypropylene (PP), are commonly used in this technology [1].

One of the crucial limitations in the development of this technology is the challenge of achieving proper adhesion between the filaments and the layers, as well as preventing the formation of voids within the structure of the parts. This issue arises because the diffusion coefficient of molten thermoplastics, which exhibit very high viscosity, remains low. In the absence of external force or pressure, ensuring strong adhesion between each melted filament and the previously deposited neighboring filament – already solidified due to rapid cooling – poses a significant challenge [2-4].

During the transformation of polymers, several parameters can influence the performance of the final part. Notably, viscoelasticity (a combination

Industrial Strategies and Solutions for 3D Printing: Applications and Optimization, First Edition. Edited by Hamid Reza Vanaei, Sofiane Khelladi, and Abbas Tcharkhtchi.

of viscosity and modulus) and the flow rate play significant roles. These parameters are inherently connected to the rheological properties of the polymer. Consequently, studying the rheology of the polymer during FFF is crucial for comprehending the underlying phenomena and optimizing the manufacturing process [5].

The rheological properties of the polymer exhibit high sensitivity to temperature variations, leading to changes in its physical state from solid to molten and vice versa. As the polymer undergoes these transitions, its rheological characteristics also undergo significant shifts. Consequently, temperature stands out as a major influential parameter throughout the FFF process. Moreover, a polymer's rheological attributes are further influenced by the rate of strain or flow. Given the correlation between rheological properties and temperature, it becomes reasonable to establish a relationship between these properties and the evolving temperature of the deposited filaments during the FFF process. This concept facilitates the creation of a connection between rheological properties, such as viscosity – serving as an indicator of the "transformation" – and the "temperature" of filaments over "time" during the deposition process. This integration leads to the development of a transformation-temperature-time (TTT) diagram. The TTT diagram provides invaluable insights into the optimal temperature and viscosity conditions necessary for the process. Thus, the TTT diagram assumes a pivotal role in optimizing the viscosity–temperature interplay during specific deposition periods [6, 7].

The importance of rheology and how it should be considered in the development of 3D printing process and optimization objectives are discussed in the following sections.

9.2 Fundamentals of Viscosity

Rheology (derived from the Greek words "rheo," meaning "to flow," and "logos," meaning "study") is the scientific discipline that explores the deformation and flow of materials in response to applied stress. Positioned within the realm of mechanics, rheology examines the interplay between a material's viscosity, plasticity, and elasticity, along with its responses to pressure-induced effects. The primary aim of rheology is to dissect the mechanical behaviors of substances and establish the governing principles of their responses. Rheology draws from a range of fundamental disciplines, including material resistance, fluid mechanics, plasticity, and more. Within rheology, the terms "deformation" and "flow" carry equivalent meanings, with "deformation" applied to solids, and "flow" generally used in reference to liquids. When a material experiences external forces leading to deformation and flow, it counteracts this stress by utilizing two key physical attributes: modulus and viscosity [8].

The modulus, which signifies the material's stiffness, represents an innate resistance that remains independent of time (measured in units of Pa). On the contrary, viscosity (measured in units of Pa S) represents a time-dependent resistance. Both modulus and viscosity originate from the interactive forces among molecules, commonly known as Van der Waals forces. When the adhesion force between molecules is substantial, the resulting modulus and viscosity are correspondingly high.

In a general sense, the behavior of materials during flow and deformation can be categorized into two extreme cases. In the first scenario, a highly elastic solid relies solely on its modulus to resist stress. Here, viscosity does not factor in, and the material is purely elastic. On the opposite end, the material adopts an extensively fluid state with a negligible modulus. In this scenario, resistance to flow is predominantly governed by the low viscosity of the fluid itself, and this is characterized as a viscous fluid. As we transition from elastic solids to viscous liquids, we first encounter viscoelastic solids, where viscosity starts to play a role although it is not yet predominant. Subsequently, we come across viscoelastic liquids. In this scenario, the importance of modulus diminishes, and viscosity takes on a predominant role.

Throughout the FFF process, similar to other polymer processing methods, the material undergoes a transition in the physical state – initially from solid to liquid during extrusion, and subsequently from liquid to solid following filament deposition. Consequently, the significance of modulus and viscosity evolves during the process. While the polymer is in the solid state, the role of modulus takes precedence. However, when the polymer transitions into the liquid state, viscosity assumes a crucial role [9].

9.3 Resistance of Materials to Flow

9.3.1 Modulus

The elastic modulus stands as an inherent physical property of a material, defined as the ratio of stress to the resultant elastic deformation produced by that stress. This property correlates with the adhesive strength among molecules. A higher intermolecular force corresponds to a greater modulus, while materials with weaker intermolecular bonds tend to have lower moduli. The modulus is measured in pascals (Pa), with units of N/m^2. Effectively, the modulus represents the material's resistance to deformation when subjected to applied stress at any given moment [10].

Typically, under small strains, the stress (σ) is directly proportional to the strain (ε), and this relationship is governed by Hooke's Law, $\sigma = E\varepsilon$, where E represents the elastic modulus. When a material's deformation instantly corresponds to the applied stress, it classifies as an elastic material. In such instances, no relaxation occurs due to the imposed deformation. The simplest representation of an elastic solid is provided by the Hookean model, which can be metaphorically likened to a spring.

9.3.2 Viscosity

Viscosity is a physical property that shows the material's resistance to deformation and flow. The unit of viscosity in the international system is Pa S. It is a resistance during the solicitation time. Indeed, during the flow of the fluid, molecules, and particles are forced to slide over each other. In this situation, the internal friction between them creates resistance to flow. This resistance is all the higher as the components of the fluid are larger. Polymers have a high melt viscosity because the size of their molecules is very large, about 50 nm, compared to small molecules such as solvent molecules with a size of about 0.5 nm. In the case of polymers in the molten state, the viscosity is directly dependent on the molecular weight. Viscosity is therefore a measure of the internal friction of the liquid. So viscosity is a measure of the internal friction of the liquid. To determine it, a measurable stress is applied to the liquid, and the strain rate is measured at the same time. If the liquid has Newtonian behavior, the viscosity, h, is a constant value determined by the ratio between the imposed stress, t, and the strain rate, $d\gamma/dt$. The behavior of a viscous fluid can be modeled by a dashpot (piston), using Newton's law.

Viscosity is a very important parameter in the study of polymer flow. Its value (at molten state) is about 10^6–10^8 times that of water. In general, the viscosity is a function of certain parameters, like molecular weight of the polymer, free volume fraction, temperature, pressure, and shear rate:

- When the molecular weight is lower than the critical molecular weight Mc, corresponding to the appearance of entanglements, the viscosity of the polymers varies almost proportionally to the molecular weight. When the molecular weight is greater than Mc, the relationship between viscosity and molecular weight obeys a power law ($h = \mathrm{K\,M}^a$, with $a = 3.4$).
- Free volume in polymers corresponds to the part of the volume not occupied by macromolecules. It is a part of volume in which the macromolecules, depending on the physical state of the polymer, can have their movements. This part of the volume is represented by a coefficient called free volume fraction, f, which can influence viscosity according to free volume theory. The relationship between viscosity and volume is

expressed by Doolittle's equation: $\eta = A \exp(B/f)$ where A et B are the constants.

- Temperature has an important influence on viscosity. If it increases, the viscosity decreases. In general, the viscosity of polymers in the liquid state varies in function of temperature, according to an exponential law (Arrhenius):
  $\eta = \eta_0 \exp(E_a/RT)$, where E_a is the activation energy, and R is the rare gas constant.
- In glass transition zone, the influence of temperature on viscosity is explained by WLF (Williams, Landel et Ferry) [11]:

$$\log \frac{\eta}{\eta_{ref}} = \frac{-C_1(T - T_{ref})}{C_2 + (T - T_{ref})}$$

- where η_{ref} is the viscosity at T_{ref} which is taken as a reference temperature. C_1 and C_2 are the constants.
- The effect of pressure on viscosity is opposite to the effect of temperature. By increasing the pressure, the viscosity also increases. The coupling effect of pressure and temperature may be shown by the following equation:
  $\eta(T, P) = A\, e^{B/T}\, e^{\Gamma P}$
- *with* Γ *coefficient dependence on* pressure:

$$\Gamma = \frac{d \ln \eta}{dP} \qquad \Gamma = 2 \times 10^{-8} - 6 \times 10^{-8}\,\text{Pa}^{-1}$$

- Strain rate, when high, also has a significant effect on the viscosity of polymer melts.

  This influence can be shown by a power law model as follows [12]:

  $$\eta = m\, \dot{\gamma}^{n-1}$$

  with
  m, consistency index, is a function of T

  $$m = m_o e^{(-b(T-T_o))}$$

  m_0 is the consistency index at the reference temperature T_0.
  B is a value between 0.01 and 0.1 °C^{-1}.
  n, index of the power law (dimensionless).

9.3.3 Relaxation Time

In the realm of materials science, relaxation time refers to a crucial concept that illuminates the response of materials to external influences over time.

This phenomenon is particularly relevant in polymers, glasses, and other viscoelastic materials. Relaxation time characterizes the duration it takes for a material to adjust or relax its internal structure and properties in reaction to a change in an external force or perturbation. Viscoelastic materials, which encompass a broad range of substances including polymers and certain biological tissues, exhibit both viscous and elastic properties. When subjected to stress, these materials not only deform elastically but also experience viscous flow over time. This dual behavior leads to a time-dependent response that can be characterized through relaxation time. Essentially, relaxation time acts as a bridge between the material's immediate elastic response and its more gradual viscous flow [13].

In polymer processing techniques such as extrusion and molding, understanding relaxation time is paramount. During these processes, polymers experience deformation and flow, causing their molecular chains to reconfigure. The relaxation time helps determine the rate at which this reconfiguration occurs. Short relaxation times signify rapid adjustment to external forces, while longer relaxation times denote a slower, more sustained response. This knowledge aids in optimizing processing conditions to achieve desired material properties and product outcomes.

Relaxation time is closely tied to rheological behavior, where the response of materials to applied stress is studied. In rheological measurements, relaxation time sheds light on how materials transition from their initial state to a more relaxed state after being subjected to stress. This information is pivotal in fields such as polymer engineering, where tailoring material properties to specific applications requires a deep understanding of relaxation behavior [14].

9.4 Materials with Different Rheological Behaviors

The rheological behavior of materials depends very strongly on the type of force imposed, the temperature, and the speed (or frequency) of stress. For this reason, the behavior of a material is not always stable. It varies under the effect of different parameters and during mechanical and thermal stresses. For example, a polymer which can exhibit a glassy behavior at low temperatures (or at high frequency), can have a rubbery behavior at higher temperatures; and if the temperature increases further, it can behave like a viscous liquid. However, under the same conditions (temperature and frequency), the different materials can be classified according to their behavior.

9.4.1 Elastic Materials

In this case, we can distinguish materials with the following different behaviors:

(i) **Ideal Solid**: This type of material is non-deformable and there is no deformation, $\varepsilon = 0$.
(ii) **Perfectly Elastic Solid**: For this type of material, there is a direct relationship between stress and strain, $\sigma = f(\varepsilon)$. When this relation is linear ($\sigma = E \times \varepsilon$), the material is a Hookean solid.
(iii) **Solid with Delayed Elastic Behavior**: For this type of solid, the deformation is a function of stress and time, $\varepsilon = f(\sigma, t)$.

The behavior of this type of material can be modelized by Voigt–Kelvin model, schematized by a spring and a damper in parallel.

$$\gamma = \gamma_{\text{Spring}} = \gamma_{\text{Piston}}$$

$$\tau = \tau_{\text{Spring}} + \tau_{\text{Piston}}$$

The model is then:

$$\tau = G\gamma + \eta\dot{\gamma}$$

The evolution of strain can have the following model:

$$\varepsilon = \lambda \times J \int_{-\infty}^{t} [\exp(-\lambda (t - \tau))] \times \sigma(\tau)\, d\tau$$

9.4.2 Viscous Materials

In terms of behavior, these materials can be presented in the following three categories.

(i) **Perfectly Mobile Liquid**: This material can be in motion without requiring constraint: $\sigma = 0$.
(ii) **Viscous Liquid**: In the case of this material, the stress is a function of the strain rate. $\sigma = f(d\varepsilon/dt)$. If this relation is linear ($\sigma = \eta \cdot d\varepsilon/dt$) the behavior is Newtonian.
(iii) **Viscoelastic Material**: This type of material exhibits a mixture of two types of elastic and inelastic behavior. In this case, during the creep test, when the constant stress is suddenly eliminated, if the deformation is greater than the elastic limit, the specimen does not return to its initial state. A residual strain remains in the sample.

The behavior of this type of material can be modelized by Maxwell's model, in which a spring and a dashpot are in series.

In this case:

$$\gamma = \gamma_{\text{piston}} + \gamma_{\text{spring}}$$
$$\dot{\gamma} = \dot{\gamma}_{\text{piston}} + \dot{\gamma}_{\text{spring}}$$
$$\dot{\gamma} = \frac{\tau}{\eta} + \frac{\dot{\tau}}{G}$$
$$\tau + \frac{\eta}{G}\dot{\tau} = \eta\dot{\gamma}$$

Then:

$$\tau + \lambda\dot{\tau} = \eta\,\dot{\gamma}$$

With: $\lambda = \eta\,/G$ relaxation time

9.4.3 Plastic Materials

According to their behavior, there are three categories of plastic materials.

(i) **Perfectly Plastic Inelastic Solid**: For this solid, there is a threshold stress, σ_{c}.
  (a) For $\sigma < \sigma_{\text{c}}$, there is no deformation.
  (b) At σ_{c}, there is a deformation without stress variation.
  (c) For $\sigma > \sigma_{\text{c}}$, the strain is permanent.
  (d) In case of discharge, the deformation is perfectly residual.

(ii) **Perfectly Plasto-elastic Material**: In this type of material, there is also a threshold stress σ_{c}.
  (a) For $\sigma < \sigma_{\text{c}}$, there is an elastic deformation.
  (b) At σ_{c}, there is a deformation without variation of the stress.
  (c) For $\sigma > \sigma_{\text{c}}$, the deformation is composed of two parts: elastic part and plastic part.
  (d) In case of unloading, the material recovers the elastic part of the deformation, but the plastic part of the deformation remains in it.

(iii) **Elasto-Viscoplastic Material**: The deformation is a function of the threshold stress and the time.

$$\varepsilon = f\,(\sigma_{\text{c}}, t)$$

  (a) If $\sigma < \sigma_{\text{c}}$, the behavior is purely elastic.
  (b) If $\sigma > \sigma_{\text{c}}$, there is viscoelastic behavior and flow of the material.

Given the abovementioned models and behaviors, the related schematics are presented in Table 9.1.

Table 9.1 The signification of different models and behaviors in viscosity evaluation [15, 16].

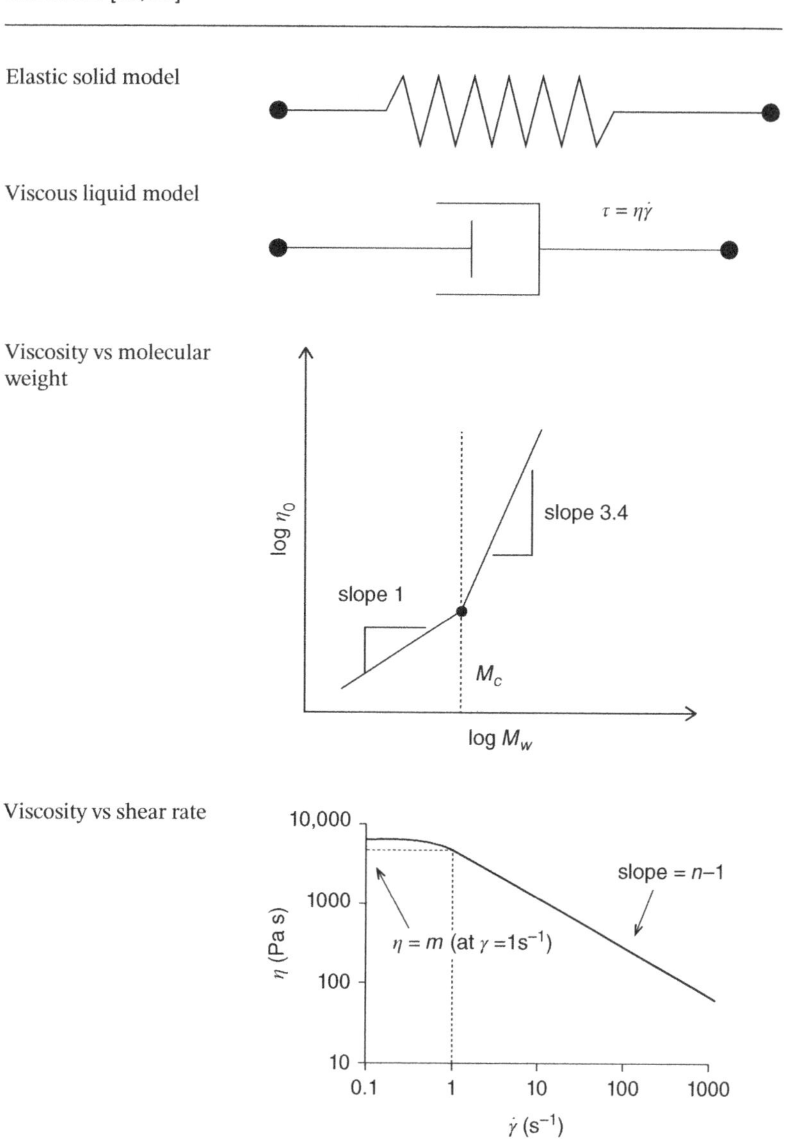

(Continued)

Table 9.1 (Continued)

Perfect elastic solid (nonlinear)	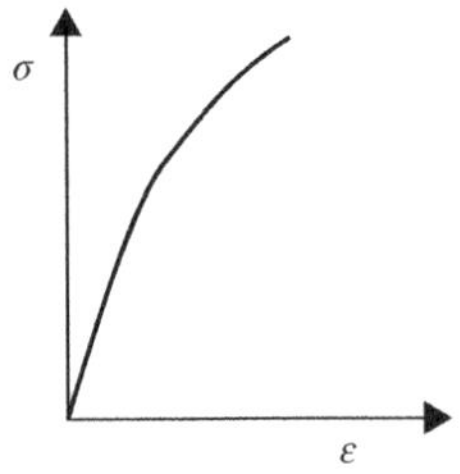 σ, ε
Perfect elastic solid (Hookean)	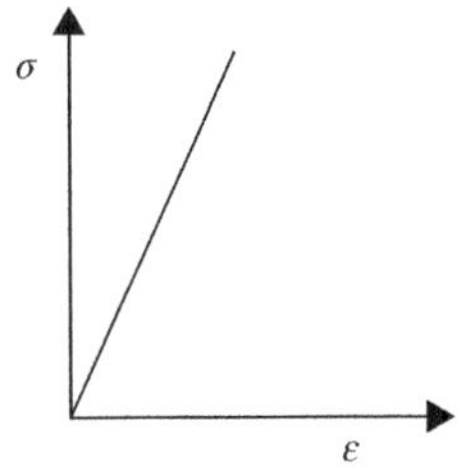 σ, ε
Voigt–Kelvin model	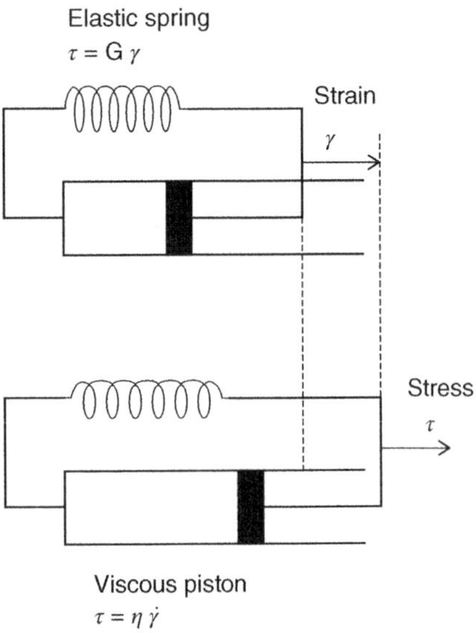 Elastic spring $\tau = G\,\gamma$; Strain γ; Stress τ; Viscous piston $\tau = \eta\,\dot{\gamma}$

(Continued)

Table 9.1 (Continued)

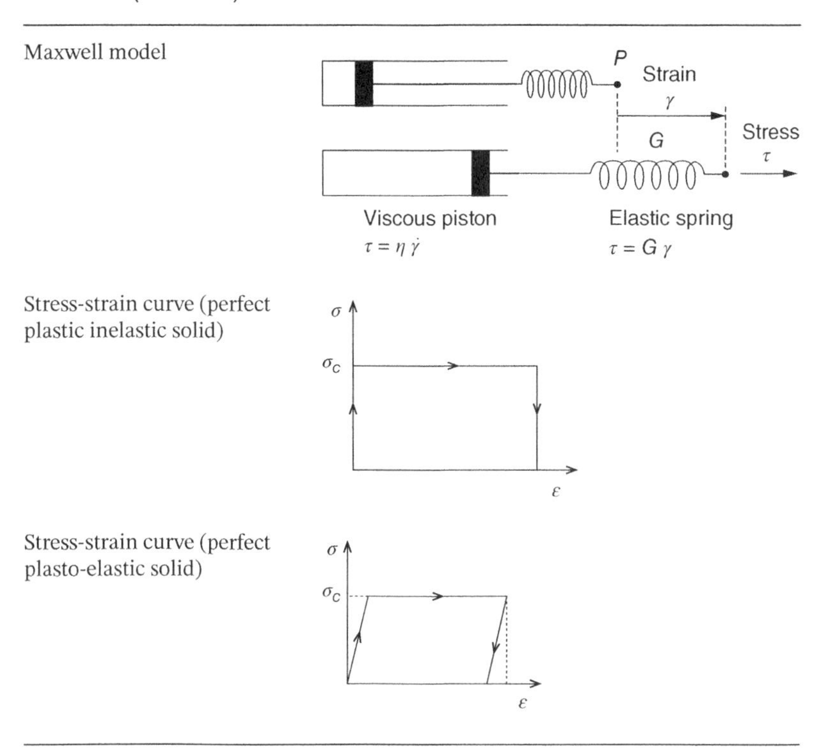

Model	Diagram
Maxwell model	
Stress-strain curve (perfect plastic inelastic solid)	
Stress-strain curve (perfect plasto-elastic solid)	

9.5 Different Rheological Behaviors at Constant Pressure and Temperature

9.5.1 Newtonian Liquids

For Newtonian liquids, the viscosity is independent of the applied stress and strain rate. Perfectly mobile liquids and viscous liquids, when the stress is a linear function of the strain rate, are the Newtonian liquids: $\sigma = \eta \cdot d\varepsilon/dt$ [17].

9.5.2 Time-independent Non-Newtonian Liquids

Three non-Newtonian liquid types can be distinguished [18]:

Shear-Thinning Liquids: In this case, the viscosity is not constant. It decreases with stress and strain rate. The majority of thermoplastic polymers in their molten state have a shear-thinning behavior.

Shear-Thickening Liquids: The viscosity of this type of liquid increases with an increase in stress or strain rate. This behavior can be observed in the case of reactive systems – cross-linking of thermosets, and polymerization of thermoplastic.

Plastic Bodies: This type of material does not flow to a stress, sc. We will have the flow for stresses greater than sc. An example of this type of material is paints.

9.6 Viscoelastic Behavior

Viscoelastic behavior is a combination of elastic and viscous effects. It is recalled that the elasticity of a material is a property linked to the rigidity of the material and reflects its capacity to conserve and restore its energy after deformation. The viscosity of a material is a property that reflects its resistance to flow (deformation) and reflects its ability to dissipate energy. The viscoelastic behavior in the solid state is dominated by the elastic behavior and, in the liquid state, by the viscous behavior. In other words, the behavior of a viscoelastic material is a behavior between that of an ideal elastic solid of modulus E (or G) and that of a Newtonian viscous liquid of viscosity η. The elasticity of a material is a property related to the stiffness of the material and reflects its ability to retain and release energy after deformation. The viscosity of a material is a property that reflects its resistance to flow (deformation) and reflects its ability to dissipate energy. This is the case with polymers where, depending on the time scale of the test, the elastic component or the viscous component of the material dominates. Indeed, for a very short stress duration (very fast stress), a linear amorphous polymer adopts a glassy (elastic) behavior. Conversely, for a very long duration of application of the stress (creep test), it exhibits viscous behavior with the possibility of flow. Polymers are macromolecular materials of very high molecular weight. At temperatures below the glass transition temperature, T_g, the amorphous phase is in the glassy state. In this state, chain movements are very limited, and an amorphous polymer essentially behaves like an elastic material. At a temperature sufficiently above T_g, when the amorphous phase is in the rubbery state, segmental movements will occur, and the polymer will lose its rigidity. In this case, the polymer acts as a vicious material. This is why for amorphous polymers, the glass transition zone is considered as the viscoelastic zone.

Viscoelastic materials can exhibit linear or nonlinear behavior. To show linear viscoelastic behavior, the following two conditions must be met – linear elasticity (a linear relationship between stress and strain), and linear viscous behavior (a linear relationship between stress and strain rate). In other words, the linear viscoelastic behavior is intermediate between the Hookean elastic behavior and the Newtonian viscous behavior. When stress and strain and/or stress and strain rate are not proportional, then the viscoelasticity is nonlinear [19].

The viscoelastic behavior of molten polymers at large deformations can induce a set of specific phenomena, such as:

- **Swelling at the Die Outlet**: One can describe the phenomenon of swelling at the exit of die of an extruder in the following way – a molten thermoplastic polymer extruded through a die retains the memory, linked to the viscoelastic behavior of the polymer, of the deformations undergone during the passage of the contraction. The material recovers its non-contracted state and swells just at the exit of the die (Figure 9.1a).
- **Weissenberg Effect**: When a cylindrical rod is rotated in a container containing the molten polymer, an elevation of the material along the rod is observed (Figure 9.1b). This effect was explained by Weissenberg in 1947. The high viscosity creates shear between the different layers of the polymer and between the polymer and the surface of the rod. This allows the polymer to rise and stay on the surface of the rod.
- **Siphon Without Tube**: When siphoning a liquid and the siphon tube is no longer in contact with the liquid, for a viscous fluid, the flow stops, while for a visco-elastic (polymeric) fluid, it can continue. Here, again, the high viscosity of the molten polymer is the cause of the creation of a shear stress between the different layers and between the polymer and the internal surface of the tube (Figure 9.1c).

As explained before, the rheological behavior can be modeled by the combination of springs and pistons. The simplest models are the Maxwell

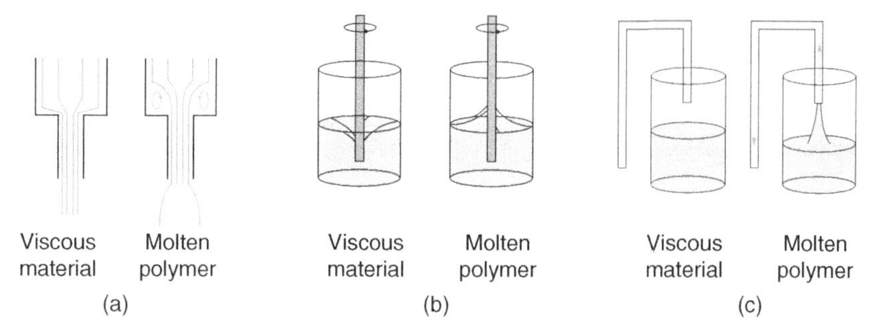

Figure 9.1 (a) Swelling, (b) Weissenberg, and (c) Siphon without tube phenomena of viscoelastic polymers.

models (a spring and a piston in series), and the Voigt–Kelvin model (a spring and a piston in parallel). However, these models are not able to express the rheological behavior of the polymer in a period of significant stress. For this reason, models of several elements are proposed in the literature.

9.7 3D Printing of Thermoplastic Polymers

The behavior and properties of viscoelastic materials play a crucial role in 3D printing processes, FFF method. FFF allows for the fabrication of 3D-printed structures using thermoplastic polymers with enough fluidity in molten state to be easily extruded. This process involves depositing molten filaments through a nozzle at a controlled flow rate over a surface, building the object layer-by-layer. The choice of thermoplastic polymer is essential to achieve complex 3D-printed structures, and several factors must be considered during the FFF process. Figure 9.2 shows some of these factors which are linked to the nature of the polymer [20].

Among these parameters related to the nature of the polymer, the role of the rheological property is the main one. Tailoring the rheology of the polymer is necessary to maintain the shape and structural integrity of the objects. Apart from the nature of the polymer, the FFF process also depends on all the different parameters such as temperature, nozzle diameter, sweep speed, layer thickness, and frame angle. Among these parameters, in this chapter,

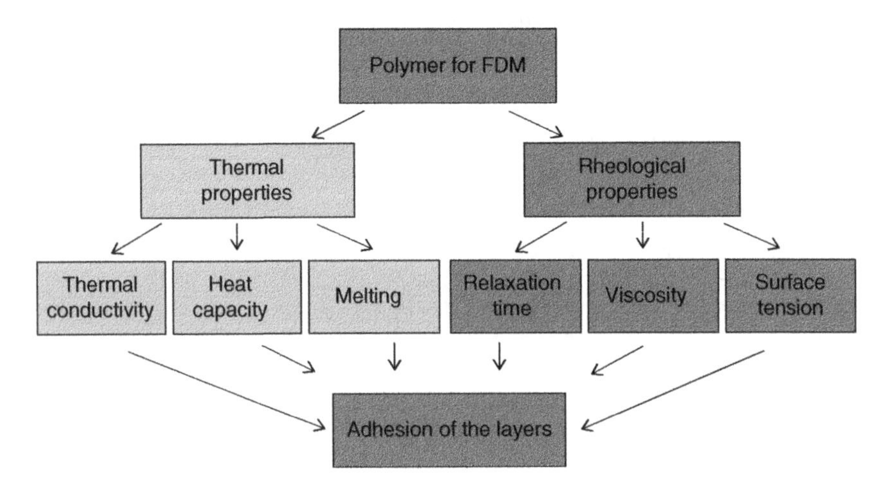

Figure 9.2 Different physical properties of the polymer having an influence on the FFF process.

we are particularly interested in the role of the rheological property of the polymer during the FFF process. However, as this property itself is influenced by temperature, we first briefly study the evolution of temperature.

9.7.1 Temperature Evolution as an Indicator for Viscosity Measurement

One of the performance criteria for parts made by the FFF process is the strength of the bonds formed between the filaments. The larger the size of the interphase between the filaments, the stronger the bonding of the deposited filaments and therefore the better the part performs. Hence, the thickness of this interphase is a function of temperature profile of filaments. Following the temperature evolution of a filament during the FFF process gives the possibility to follow its viscoelasticity evolution, hence the bond and interphase formation between the filaments. In the case of PLA, this temperature evolution during the process was followed using very thin (50 μm diameter) thermocouples (Figure 9.3). This variation in temperature of each filament corresponds to the thermal effect of the following successive melted filaments. The temperature of the already deposited filament increases due to contact with the next hot filament; then it decreases.

This cycle continues several times. However, each time the intensity of the cycle will be less. In this diagram, T_c and T_m correspond respectively to the crystallization temperature and to the melting point of the polymer. For good adhesion between the filaments, they must always remain in the molten state at the time of coalescence with the following filament, that is to say, at a temperature above the crystallization temperature, T_c. At a temperature below T_c, the polymer is already solidified and its coalescence with the next molten filament is practically impossible. The evolution of the temperature of filaments during FFF process can be modelized, using the heat transfer equation (Figure 9.4).

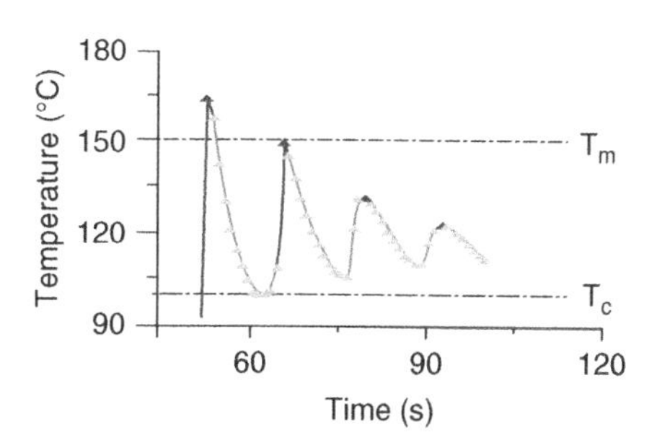

Figure 9.3 Temperature evolution of a filament during the FFF process.

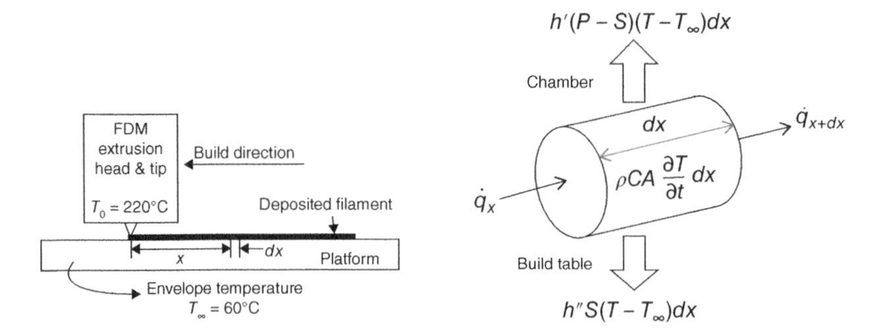

Figure 9.4 Heat transfer of filament during FFF process.

The modeling has been based on the following assumptions:

- The polymer melt is assumed to have a circular cross-section.
- The temperature is constant across the cylindrical rod.
- The filament is deposited at a constant rate.
- The conduction heat transfer between the filament and build table is assumed to be in the form of convection for simplifying the calculation.

The heat transfer equations are as follows [21]:

- Energy out from the right surface $= \dot{q}_{x+dx} = \dot{q}_x + \frac{\partial \dot{q}_x}{\partial x} dx = -A\left[k\frac{\partial T}{\partial x} + \frac{\partial\left(k\frac{\partial T}{\partial x}\right)}{\partial x} dx\right]$
- Convection heat transfer with air $= h'(P - S)(T - T_\infty)dx$
- Convection heat transfer with build table $= h''S(T - T_\infty)dx$
- Internal energy change $= mC\frac{\partial T}{\partial t} = \rho CA\frac{\partial T}{\partial t}dx$

So: Energy in – Energy out = Change in internal energy

$$A\frac{\partial\left(k\frac{\partial T}{\partial x}\right)}{\partial x} - hP(T - T_\infty) = \rho CA\frac{\partial T}{\partial t}$$

In addition: $X = v * t$ with $\frac{\partial x}{\partial t} = v$

So: $\frac{\partial T}{\partial t} = \frac{\partial x}{\partial t} \times \frac{\partial T}{\partial x}\frac{\partial T}{\partial t} = v \times \frac{\partial T}{\partial x}$

Therefore, we will have:

$$A\frac{\partial\left(k\frac{\partial T}{\partial x}\right)}{\partial x} - hP(T - T_\infty) = \rho CAv\frac{\partial T}{\partial x}$$

This is the fundamental equation for energy balance, in which relates the temperature of the element with respect to the position of the nozzle. With the following boundary conditions:

$$T = T_0 \quad \text{at} \quad x = 0\,(\text{Nozzle tip}) \quad \text{at} \quad t \geq 0$$

$$T = T_\infty \quad \text{at} \quad x = \infty \quad \text{at} \quad t \geq 0$$

By solving the equation using the above boundary condition and considering the lumped capacity for modeling the cooling process of the extruded filament, we will have:

$$T = T_\infty + (T_0 - T_\infty)e^{-mx}$$

where:

$$m = \frac{\sqrt{1 + 4\alpha\beta} - 1}{2\alpha}$$

$$\alpha = \frac{k}{\rho Cv}$$

$$\beta = \frac{hP}{\rho CAv}$$

The signification of different parameters used in the above equations is presented in Table 9.2.

Given the abovementioned explanations, temperature evolution serves as a valuable indicator for viscosity measurement, offering insights into the

Table 9.2 The signification of different parameters.

A	Cross-section area of elliptical filament
P	Perimeter of elliptical cross-section
h'	Convection coefficient between filament and air
h''	Convection coefficient between filament and build table
h	Effective convection coefficient
k	Thermal conductivity of filament
m	Mass of filament
C	Specific heat of filament
ρ	Density of filament
S	Contact length between filament and build table
T_∞	Envelope/build table temperature
T_0	Extrusion temperature
v	Velocity of the extrusion head
x	Nozzle travel distance in time "t"

flow behavior and consistency of materials. As temperature changes, the viscosity of many substances can experience significant alterations due to their temperature-sensitive molecular interactions. By observing how the viscosity of a material changes as it is subjected to varying temperatures, researchers and engineers can gain a deeper understanding of its behavior during different processing or application conditions. The relationship between temperature and viscosity is often governed by well-known models, such as the Arrhenius equation, which can provide valuable predictions about how viscosity will behave at different temperatures. This information is particularly crucial in industries such as manufacturing, where the viscosity of materials like polymers, paints, and adhesives can greatly influence their processability and end-product quality. By utilizing temperature as an indicator for viscosity measurement, professionals can make informed decisions in optimizing manufacturing processes, ensuring product consistency, and achieving desired material performance across diverse environmental conditions.

9.7.2 Interphase Formation Between the Filaments During 3D Printing Process

Interphase formation between filaments during the 3D printing process plays a crucial role in determining the overall structural integrity and mechanical properties of the final printed object. This interphase, often referred to as the "interlayer bond," develops at the boundary where two adjacent filaments merge together during the printing process. This phenomenon occurs due to the localized heat and pressure applied during the fusion of layers, creating a region where the two filaments partially melt and fuse. This interface forms a bridge that connects the individual printed layers, effectively creating a cohesive and continuous structure. The quality and characteristics of this interphase have a direct impact on the mechanical strength and performance of the printed part. A strong and well-formed interphase ensures effective load transfer between layers, enhancing the overall structural integrity of the object. This becomes especially critical in applications where the printed object is subjected to mechanical stresses or external forces.

The formation of the interphase is influenced by several factors, including the printing material properties, printing parameters such as temperature and print speed, and the printer's design. Materials with good compatibility and similar melting temperatures tend to form more robust interphases, promoting efficient bonding between adjacent layers. However, the interphase must strike a balance between being strong enough to provide structural stability while not being too brittle, as excessive brittleness could lead to interlayer cracking or delamination.

Manufacturers and researchers often focus on optimizing the interphase to achieve the desired mechanical performance. Through precise control of printing parameters, engineers can fine-tune the temperature profiles and cooling rates to encourage proper interlayer bonding without causing deformation or defects. Additionally, advancements in multi-material and multi-nozzle printing techniques offer the ability to create gradient interphases, tailoring the properties across the interface for enhanced performance. Understanding and controlling the interphase formation is an ongoing area of research in AM. As 3D printing continues to evolve, the ability to engineer the interphase will contribute to producing more reliable and high-performance printed objects across a wide range of applications, from aerospace components to medical implants and beyond [22].

9.8 Rheology and Optimization in 3D Printing Process

Rheology plays a pivotal role in the optimization of the 3D printing process. In 3D printing, where materials are deposited layer by layer to create intricate structures, understanding the rheological properties of the printing material is essential for achieving accurate and consistent prints. The viscosity, elasticity, and other rheological characteristics of the material dictate how it will behave during extrusion, deposition, and solidification.

Optimizing the 3D printing process involves a delicate balance of various parameters, and rheology provides insights into how these parameters interact. By comprehending the flow behavior of the material at different temperatures, shear rates, and pressures, engineers can fine-tune printer settings to ensure proper material deposition and layer adhesion. For instance, adjusting the printing speed, nozzle size, and temperature can influence the material's viscosity during extrusion, impacting its flow rate and subsequent layer bonding. Moreover, rheological analysis aids in material selection and formulation. Different materials, such as thermoplastics, metals, ceramics, and biomaterials, exhibit diverse rheological behaviors that must be compatible with the printing process. The formulation of printing materials involves optimizing their rheological properties to enable smooth extrusion, minimize sagging, and prevent deformation during printing. A thorough understanding of rheology can lead to the development of tailor-made materials optimized for 3D printing applications.

Rheological measurements also offer feedback for quality control and process consistency. Monitoring changes in the material's viscosity during the printing process can help detect potential issues like clogging, inconsistent flow, or poor layer bonding. Real-time rheological data can be integrated into closed-loop control systems, allowing printers to adjust parameters on-the-fly for optimal print quality. Overall, rheology is a cornerstone in the quest for optimizing the 3D printing process. It empowers engineers and researchers to understand, manipulate, and predict how materials will behave during printing, thereby enabling the production of high-quality, accurate, and structurally sound 3D-printed objects across a wide range of applications [23].

9.9 Summary

Rheological evaluation has emerged as a critical facet in the realm of 3D printing, playing a pivotal role in enhancing the precision, efficiency, and quality of the AM process. As 3D printing continues to expand across various industries, the study of how materials flow and deform has gained prominence for its capacity to influence material behavior during deposition, layer bonding, and solidification. The ability to comprehend and manipulate the rheological characteristics of printing materials is foundational for producing consistent and intricate objects. Understanding parameters like viscosity, elasticity, shear thinning, and yield stress facilitates the selection and formulation of materials that align with the printing process. Tailoring rheology to the specific requirements of a 3D printing application, be it high-resolution prototypes or functional end-products, ensures that materials can be smoothly extruded, deposited with precision, and bonded securely between layers.

Rheological evaluation also contributes to the optimization of the 3D printing process itself. By studying how materials respond to temperature changes, shear forces, and extrusion rates, engineers can fine-tune printer settings to achieve optimal results. Such insights enable adjustments to printing speed, temperature profiles, and nozzle size, ultimately influencing material flow, layer adhesion, and overall structural integrity. The integration of real-time rheological data into closed-loop control systems enhances process reliability and consistency, as printers can adapt on-the-fly to changes in material behavior. Furthermore, rheological evaluation aids in diagnosing and troubleshooting issues during printing. Monitoring shifts in viscosity and other rheological properties offers insights into potential challenges like clogging, inconsistent flow, or poor bonding between layers. This early detection capability enables swift adjustments, preventing costly production failures and minimizing material waste.

Rheological evaluation stands as an indispensable tool in the optimization of 3D printing processes. Its impact spans material selection, formulation, process refinement, and quality control, all of which contribute to the creation of high-quality, precisely fabricated 3D-printed objects. As AM continues to evolve, rheological insights will undoubtedly remain instrumental in pushing the boundaries of design, functionality, and performance in 3D printing across diverse industries.

References

1 Yang, K., Grant, J.C., Lamey, P. et al. (2017). Diels–Alder reversible thermoset 3D printing: isotropic thermoset polymers via fused filament fabrication. *Advanced Functional Materials* 27 (24): 1700318.

2 Kantaros, A. and Piromalis, D. (2021). *Employing a low-cost desktop 3D printer: challenges, and how to overcome them by tuning key process parameters. International Journal of Mechanics and Applications* 10 (1): 11–19.

3 Vaudreuil, S., Bencaid, S.E., Vanaei, H.R. et al. (2022). *Effects of power and laser speed on the mechanical properties of AlSi7Mg0. 6 manufactured by laser powder bed fusion. Materials* 15 (23): 8640.

4 Vanaei, H.R., Khelladi, S., and Tcharkhtchi, A. (2022). Roadmap: numerical-experimental investigation and optimization of 3D-printed parts using response surface methodology. *Materials* 15 (20): 7193.

5 Vanaei, H.R., Magri, A.E., Rastak, M.A. et al. (2022). *Numerical–experimental analysis toward the strain rate sensitivity of 3D-printed nylon reinforced by short carbon fiber. Materials* 15 (24): 8722.

6 Khaliq, M.H., Gomes, R., Fernandes, C. et al. (2017). *On the use of high viscosity polymers in the fused filament fabrication process. Rapid Prototyping Journal* 23 (4): 727–735.

7 Bom, S., Ribeiro, R., Ribeiro, H.M. et al. (2022). *On the progress of hydrogel-based 3D printing: correlating rheological properties with printing behaviour. International Journal of Pharmaceutics* 615: 121506.

8 Meng, Q., Liu, H., and Wang, J. (2017). *A critical review on fundamental mechanisms of spontaneous imbibition and the impact of boundary condition, fluid viscosity and wettability. Advances in Geo-Energy Research* 1 (1): 1–17.

9 Huang, B. and Bártolo, P.J. (2018). *Rheological characterization of polymer/ceramic blends for 3D printing of bone scaffolds. Polymer Testing* 68: 365–378.

10 Zhang, J., Alvarez, A.E., Lee, S.I. et al. (2013). *Comparison of flow number, dynamic modulus, and repeated load tests for evaluation of HMA permanent deformation. Construction and Building Materials* 44: 391–398.

11 Sopade, P., Halley, P., Bhandari, B. et al. (2003). *Application of the Williams–Landel–Ferry model to the viscosity–temperature relationship of Australian honeys. Journal of Food Engineering* 56 (1): 67–75.

12 Boutelier, D., Schrank, C., and Cruden, A. (2008). *Power-law viscous materials for analogue experiments: new data on the rheology of highly-filled silicone polymers. Journal of Structural Geology* 30 (3): 341–353.

13 Rendell, R., Ngai, K., and McKenna, G. (1987). *Molecular weight and concentration dependences of the terminal relaxation time and viscosity of entangled polymer solutions. Macromolecules* 20 (9): 2250–2256.

14 Lou, N., Wang, Y., Li, X. et al. (2013). *Dielectric relaxation and rheological behavior of supramolecular polymeric liquid. Macromolecules* 46 (8): 3160–3166.

15 Osswald, T. and Rudolph, N. (2015). *Polymer Rheology*. München: Carl Hanser.

16 Afoakwa, E.O. et al. (2009). *Comparison of rheological models for determining dark chocolate viscosity. International Journal of Food Science & Technology* 44 (1): 162–167.

17 Zhu, Y. and Granick, S. (2001). *Rate-dependent slip of Newtonian liquid at smooth surfaces. Physical Review Letters* 87 (9): 096105.

18 Dziubinski, M., Fidos, H., and Sosno, M. (2004). *The flow pattern map of a two-phase non-Newtonian liquid–gas flow in the vertical pipe. International Journal of Multiphase Flow* 30 (6): 551–563.

19 Polacco, G. et al. (2006). *Relation between polymer architecture and nonlinear viscoelastic behavior of modified asphalts. Current Opinion in Colloid & Interface Science* 11 (4): 230–245.

20 Jiang, Z. et al. (2020). *Extrusion 3D printing of polymeric materials with advanced properties. Advanced Science* 7 (17): 2001379.

21 Nellis, G. and Klein, S. (2008). *Heat Transfer*. Cambridge University Press.

22 Lacambra-Andreu, X. et al. (2023). *Rheological investigation and modeling of healing properties during extrusion-based 3D printing of poly (lactic-acid). Rheologica Acta* 62 (1): 31–44.

23 Hernández-Sosa, A. et al. (2022). *Optimization of the rheological properties of self-assembled tripeptide/alginate/cellulose hydrogels for 3D printing. Polymers* 14 (11): 2229.

10

Investigating the Mechanical Performance of 3D-printed Parts

Hamid Reza Javadinejad[1], *Abdoulmajid Eslami*[1], *and Hamid Reza Vanaei*[2,3,4]

[1] *Department of Materials Engineering, Isfahan University of Technology, Isfahan, Iran*
[2] *ESILV, Léonard de Vinci Pôle Universitaire, 92916 Paris La Défense, France*
[3] *Léonard de Vinci Pôle Universitaire, Research Center, 92916 Paris La Défense, France*
[4] *Arts et Métiers Institute of Technology, CNAM, LIFSE, HESAM University, 75013 Paris, France*

10.1 Introduction

3D printing is considered a disruptive technology for the fabrication of industrial components, and thus, mechanical behaviors of 3D printing parts are very important in terms of their applications. Regarding 3D-printed polymeric materials due to their intrinsic soft nature, mechanical properties need more attention. Additionally, there are many parameters that influence mechanical behaviors of 3D-printed polymeric parts. The most important mechanical properties subjected to research are as follows [1]:

- Modula
- Tensile
- Compressive
- Flexural
- Impact
- Shear
- Hardness
- Fatigue
- Creep

There are two main standards measuring the mechanical properties of 3D-printed polymeric materials. In this regard, in the American Society for Testing Materials International (ASTM) standard, committee F42 defines additive manufacturing (AM) technologies, and subcommittee F42.05

Industrial Strategies and Solutions for 3D Printing: Applications and Optimization, First Edition. Edited by Hamid Reza Vanaei, Sofiane Khelladi, and Abbas Tcharkhtchi.

addresses materials and processes. The other standard is the International Standards Organization (ISO). Both of the standards address the mechanical testing of AM 3D-printed polymeric materials and parts [2].

In addition to many experiments which have been conducted and explored the mechanical properties of 3D-printed polymeric parts, some mathematical models have also been developed [3]. In this regard, this chapter reviews the mechanical properties of 3D-printed polymeric materials and their influencing factors.

10.2 Mechanical Properties of 3D-Printed Parts

10.2.1 Modula of 3D-Printed Parts

Mechanical properties of 3D-printed parts are mostly defined in terms of their behavior exposed to loads. For solid materials, moduli are defined as the stress divided by the strain. Moduli can be measured under different loading modes including tension, compression, flexure, shear, or torsion for polymeric materials. Moduli is a function of temperature and time, but its temperature dependency is more important than time. For many polymers used for 3D printing, these properties are often reported in technical specification sheets at different loading modes [2].

10.2.2 Tensile Properties of 3D-Printed Parts

Tensile strength as one of the most important quasi-static properties of 3D-printed polymeric materials refers to the amount of force that can be applied to an object or material before it fractures. If the force increases on an object or material, then it would go beyond the point of deformation and break, because the molecules of the material would not withstand the applied external forces. The force required to break the object is defined as material's tensile strength at break rating. The tensile strength is mostly reported in MPa or GPa and is measured as follows:

$$\sigma_{\max} = \frac{P_{\max}}{A_0} \tag{10.1}$$

where $P_{\max}$ is maximum load and A_0 is the original cross-sectional area. The tensile Young's modulus is calculated from the slope of the stress-strain curve with the stresses σ_1 and σ_2 [2]:

$$E_t = \frac{\sigma_2 - \sigma_1}{\varepsilon_2 - \varepsilon_1} \tag{10.2}$$

The technical standards for determining tensile properties for plastics are ASTM D638, ISO 527-2, and for composites, they are ASTM D3039, ISO 527-4. According to the mentioned standards, dog bone or end tab specimens can be used to determine tensile strength; however, geometry is based on the thickness of the sample or the type of material being used [2].

There has been some research investigating the tensile properties of 3D-printed joints. In this regard, Grasso et al. explored the combined effect of temperature and filament orientation on the mechanical properties of the 3D-printed PLA by FDM under different thermal conditions. According to the tensile results, a strong correlation existed between stiffness and tensile strength with the infill orientation and the temperature values [3]. Vălean et al. investigated the effects of spatial printing direction (0°, 45°, 90°) and size effect (different thicknesses) on tensile properties of 3D-printed PPLA specimens fabricated through FDM process. Their results revealed that spatial orientation has less influence on Young's modulus and higher influence on tensile strength [4]. Dwiyati et al. studied tensile properties of 3D-printed ABS material in an axial and lateral direction with different thickness layers and reported that the maximum force and tensile strength of the axial direction of 3D-printed specimens were higher than the lateral direction. The thicker layer also tended to have greater maximum force and tensile strength. In fact, more layers provide a tight structure but the strength of inner layers was lower than the strength between polymer molecules in one layer. This is because, in one layer, the polymer molecules have the same conditions of heating and cooling, so that between the polymer molecules, secondary and tertiary bonds form a stronger structure. The thicker the layer, the stronger the bond; it occurs because more polymer chains are involved in forming bonds [5]. Galeja et al. evaluated different raster angles (45, 55, 55', 60, and 90°) on the mechanical properties of 3D-printed ABS materials using FDM technique, and did not find significant changes in the tensile behaviors of 3D-printed ABS with raster angles [6].

Wang et al. proved that FDM as a thermal process may introduce internal voids and pores into the fabricated thermoplastics, leading to reduction in the mechanical properties. They explored the effects of the microscopic pores' characterizations including size, shape, density, and spatial location on the mechanical properties of material fabricated by the FDM process through experiments and micromechanical modeling. Based on the proposed micromechanical model, mechanical properties of 3D-printed materials using FDM may be a function of the porosity [7]. Garzon-Hernandez et al. predicted the mechanical behaviors of 3D-printed ABS components made by FDM through a thermal model. For modeling.

manufacturing parameters and filament properties were used as inputs, and observed that layer height, raster orientation, and number of layers have the most influence on the void density and mechanical performance of ABS components. For instance, higher elastic modulus and maximum stress were observed for lower layer heights and longitudinal raster orientation. This improvement was accompanied by a decrease in the void density. Moreover, increasing the number of layers resulted in lower mechanical properties, with the exception of the maximum tension for transverse components. This study shows the layer heights and environmental temperature as the most relevant manufacturing parameters [8].

10.2.3 Compressive Properties of 3D Printed Parts

Compressive strength refers to the capacity of a material or structure to withstand loads tending to reduce size. In a compression test, there is a linear region where the material follows Hooke's law. Hence:

$$\sigma = E\varepsilon \tag{10.3}$$

where E is Young's modulus for compression. In this region, the material deforms elastically and returns to its original length when the stress is removed. This linear region terminates at what is known as the yield point. Above this point, the material behaves plastically and will not return to its original length once the load is removed.

Generally, ASTM D695 and ISO 604 are the most popular standards for compression measurements. Additionally, ASTM D3410 and ISO 14126 are specific to compression of a fiber-reinforced composite in plane direction. The standards provide measurement of compressive modulus, compressive yield stress, compressive strength at failure, and compressive strain at failure. There are geometrical restrictions for the diameter and height of the sample in the compression test [2].

Subeshan et al. examined the influence of infill shapes (solid, diamond, hexagonal, square, and triangle), volumes (20, 40, 60, 80, and 100 vol%), and orientation of infill shapes on the compressive strength of 3D-printed PLA components. Findings indicated the infill shapes and volume percentages have the most impact on the compressive strength of 3D-printed PLA parts. Generally, high-infill shapes provide better compression strengths [9]. Nomani et al. also examined the impacts of deposition layer thickness on the mechanical properties of 3D-printed ABS material using the FFF process. Investigating both tensile and compression properties indicated that strength and stiffness were greatest using smaller layer thicknesses, compared with larger layer thicknesses. The changes in mechanical properties may be attributed to the number of deposited layers promoting interlayer bonding strength [10]. Wang et al. studied the effects of extrusion temperature and layer thickness on the quasi-static and dynamic behavior of two 3D-printed polypropylene/ethylene-propylene-rubber (PP/EPR) and polypropylene/ethylene-propylene-diene-monomer/talc (PP/EPDM) composites. They found that the extrusion process affected the compressive properties of printed specimens by inducing voids and stress concentration, resulting in crack propagation [11]. Valvez et al. discussed static behavior of 3D-printed neat PETG and PETG composites reinforced with carbon or Kevlar fibers. It was found that the compressive yield strength of the composites decreased in both composites compared to the neat polymer. Moreover, the stress relaxation behavior experienced a decrease in compressive stresses over time for neat PETG. The mechanical response was also very dependent on the displacement/stress level used at the beginning of the compression process. However, when the fibers were added to the polymer, higher stress relaxations and compressive displacements were obtained [12].

10.2.4 Flexural Properties of 3D Printed Parts

Flexural strength, also known as bend strength, is a measurement used to define a material's tendency to bend [2]. It is a ratio of stress to strain in flexural deformation. The flexural strength is widely used with materials that deform significantly but will not break. The flexural strength of a material is measured in megapascals (MPa). Thus, the flexural stress is calculated as:

$$\sigma_f = \frac{3PL}{2bd^2} \tag{10.4}$$

where P is the flexural force, L is the support span, b is the specimen width, and d is the specimen thickness. Accordingly, the flexural strength σ_f, ULT is calculated by substitution of the maximal flexural force. Then the flexural strain ε_f could be calculated as:

$$\varepsilon_f = \frac{6Dd}{L^2} \tag{10.5}$$

where D is the mid-span deflection. The flexural modulus of elasticity E_f is the ratio, within the steepest initial elastic region of stress to corresponding strain. Thus, it is calculated as:

$$E_f = \frac{L^3 m}{4bd^2} \tag{10.6}$$

where m is the slope of the secant of the flexural force-deflection curve. The slope m is determined from the linear region between 10% and 60% of the maximum flexural force.

ASTM D790 and ISO 178 are the main standards to measure the flexural modulus, flexural strength, flexural stress, and strain at break using the three-point bend method within a 5% strain limit. If the strain limit is not met, then ASTM D6272, which is a four-point method, is used to increase the chance of achieving a failure measurement. This test reduces the stress concentration associated with the center roller in the three-point test. These standards are applicable to unreinforced and reinforced materials. For composites containing high-modulus fibers, ASTM D7264 should be used for testing. It is worth mentioning that this standard does not address the specific challenges for AM materials with anisotropic properties [2].

Krzikalla et al. focused on the experimental and numerical examination of the flexural properties of long carbon fiber reinforced composites with Onyx polymeric matrix fabricated using the FDM process. Two fiber volumetric contents and two reinforced layer distributions were investigated. The results suggested that the reinforced layer distribution is crucial over the fiber content. The influence of transversal constants can be negligible under certain stress states [13]. Chen et al. (studied the flexural behavior of 3D-printed grid beetle elytron plates (GBEP), grid plate (GP), and honeycomb plate (HP) parts with the same wall thickness using three-point bending experiments. The results demonstrated that the flexural strength and energy absorption per unit mass of the GP and GBEP had significantly improved (Figure 10.1) [14]. Ibrahim et al. indicated that the wire volume

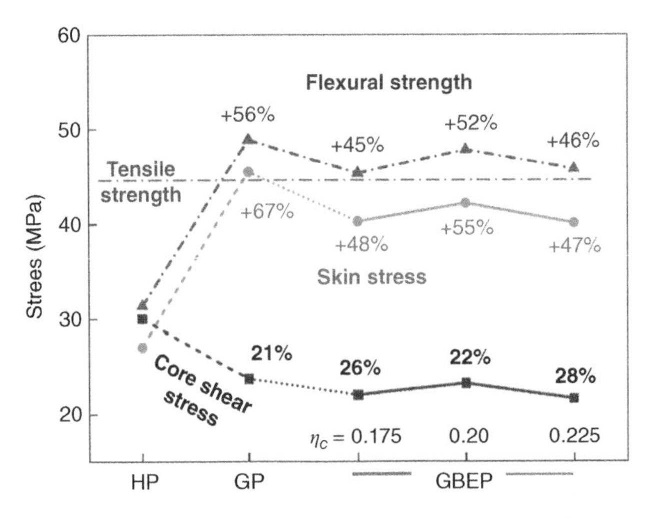

Figure 10.1 The stresses σ and τ of the weak point and flexural strength of 3D-printed GBEP, plate GP and HP plate. Source: Chen et al. [14]/Elsevier.

fraction, type of polymer matrix, and wire treatment can have positive effects on the flexural properties of 3D-printed continuous wire polymer composites (CWPCs). They reported that matrix materials, which included chopped carbon fibers and aluminum particles in conjunction with the continuous wires, provided significant improvement in the flexural strength and modulus compared to wire-reinforced PLA samples [15]. Araya-Calvo et al. reported the effect of reinforcement pattern, reinforcement distribution, print orientation, and percentage of fiber on flexural mechanical properties of polyamide 6 (PA6), reinforced with carbon fiber composite filament fabrication (CFF). The best flexural strength was achieved with 0.4893 carbon fiber volume ratio, concentric reinforcement, and perpendicular to the applied force. This may prove the effect of the 3D-printed composite response under compressive loads, in which the PA6 matrix avoids fiber bucking effects and supports the compressive stress under flexural loads; the concentration of stress is located in the continuous fibers of the 3D-printed composite [16]. Alshamrani et al. evaluated the influence of layer thicknesses (25, 50, and 100 μm) and post-processing on the mechanical properties of a 3D-printed dental resin material fabricated using digital light projection (DLP) technology. The results indicated that the higher printing layer thickness resulted in the highest flexural strength. But the printing layer thickness and post-processing treatment had no effect on the degree of conversion ratio of the 3D-printed material [17]. Li and Lou found that different printing factors including temperature, printing direction, printing path, and layer thickness have significant impacts on the tensile strength, flexural strength, crystallinity, and grain size of 3D-printed poly-ether-ether-ketone (PEEK) parts by FDM. The results demonstrated a more positive effect of higher temperatures and layer thicknesses on the mechanical performance of samples. The horizontal printing directions and the printing path 180° also showed the same influence [18].

10.2.5 Impact Strength Properties of 3D Printed Parts

The impact strength can be measured through any of the Charpy impact tests or Izod tests. The Charpy impact test makes use of a pendulum arm attached to a pre-calibrated energy gauge. The Izod impact test is similar to the Charpy test in the sense that it also uses a hammer attached to a pendulum arm to hit a custom-made specimen bar and measure the energy needed to fracture it. The main differences between the Izod test and the Charpy test are the orientation of the specimen and size, notch face direction, type of hammer, and type of tested material [19]. For a specimen, the angle β_{sample} is measured and used to calculate the uncorrected energy (E_S). The total correction energy is calculated as follows:

$$E_{TC} = \left(E_A - \left(\frac{E_B}{2}\right)\right)\left(\frac{\beta_{sample}}{\beta_{max}}\right) + \left(\frac{E_B}{2}\right) \tag{10.7}$$

The impact resistance for a specimen is calculated as:

$$I_R = \frac{E_S - E_{TC}}{t} \tag{10.8}$$

where t is the sample thickness. Similarly, the impact energy is also calculated as follows:

$$I_E = \frac{E_S - E_{TC}}{At} \tag{10.9}$$

where A is the thickness of the material under the notch [19].

ISO 179 and ASTM D6110 are specific method for the Charpy impact test, while ISO 180 and ASTM D256 describe Izod impact testing. Impact characterizations are mentioned in technical data sheets of many 3D printed relevant polymers. Both methods have widely used to tests of polymeric materials [2].

Lay et al. compared the impact strength of the 3D-printed PLA, ABS, and PA6 fabricated using FDM and injection molding. They found that the impact strength of the 3D-printed specimens was 78% lower compared to injection-molded specimens [20]. In a similar research, Patterson et al. explored the impact strength of ten different polymeric materials using seven different hatch angles and three different print orientations. They found that impact properties strongly depended on the shell orientation and raster angle, in which the crack length and path through the material during fracture can be determined [19]. Upadhyay et al. conducted the Izod impact tests on 3D-printed ABS by FDM in the horizontal and vertical orientations. They reported higher values of impact strength and hence absorbed more energy until fractures were obtained in the horizontal orientation indicated, compared to the vertical orientation [21]. Wang et al. showed the effect of layer thickness (0.2 and 0.4 mm) and table temperature (30 and 160 °C) on the Izod impact strength of 3D-printed PLA fabricated by FLM process. It was observed that the Izod impact strength of PLA fabricated at high table temperature was up to 114% higher than that of the injection-molded PLA using traditional molding parameters [22]. Kabir et al. indicated that unidirectional 3D-printed continuous glass-fiber-reinforced nylon composites had the highest impact strength, followed by cross-lay and quasi-isotropic specimens [23]. Tsou et al. observed a noticeable reduction in the impact strength of PLA and rice husk composites by increasing the rice husk content [24]. Peng et al. reported that the composite core- shell filaments provided improved impact strength compared to individual components, while the impact strength of the composite filaments containing 45% shell by volume could exceed 800 J/m [25]. Tezel et al. demonstrated the role of layer thickness (0.1, 0.2, 0.3, 0.4, 0.5 mm) and print orientation (upright, flatwise, edgewise) on the impact strength of several 3D-printed engineering polymers including ABS, PLA, glycol-modified polyethylene terephthalate (PET-G), and polycarbonate (PC) fabricated by FDM. Layer thickness (especially 0.3 mm) had the most influence on the impact strength of PC, followed by PET-G, ABS, and PLA. The lowest impact strength values were achieved for upright print orientation [26].

10.2.6 Shear Properties of 3D Printed Parts

Shear strength refers to the ability of an adhesive to resist shear stress. Shear stress is also a type of external force acting on an object or surface in a direction that is parallel to the surface. For example, the shear resistance of an adhesive is tested by bonding two surfaces together using that adhesive and then forcing one of the surfaces to slide across the other (e.g. by applying a label onto a substrate and then trying to slide the label across the surface of that substrate). The shear stress can be measured as follows:

$$\tau = \frac{\sigma_1 - \sigma_3}{2} \tag{10.10}$$

where σ is major principal stress and σ_3 is minor principal stress.

There are two specific shear standards which are directly applicable to 3D printed materials. ASTM D4255 and ISO 15310 are mostly utilized to measure the shear modulus of plastics and fiber-reinforced materials, respectively. These standards allow for testing isotropic materials, but there is no guidance for testing materials constructed via AM [2].

Cantrell et al. studied the shear characterization of 3D-printed ABS and PC parts using digital image correlation (DIC) to determine the extent of anisotropy of 3D-printed materials. Findings suggested that raster and build orientation had a positive effect on shear modulus and shear yield strength varied by up to 33% in ABS specimens. Raster orientation in the flat samples showed anisotropic behavior in PC specimens as the moduli and strengths varied by up to 20% [27]. Elmrabet et al. explored the relationship between the layer thickness, infill density, and post- processing heat-treatment time at 100 °C on the shear properties of 3D-printed PLA. It was concluded that the combination of 100% infill density and 0.1 mm layer thickness is desired to achieve the best shear properties [28]. Caminero et al. reported that weak interlaminar bonding is one of the major limitations of material extrusion-based 3D printing process, leading to shear failure of parts. They mentioned that interlaminar shear strength (ILSS) of CF, FG, and

K-reinforced isotropic nylon composites strongly depended on fiber volume fraction and layer thickness [29]. Yang et al. reported that the shear strength of 3D-printed continuous fiber-reinforced thermoplastic composites has a correlation with fiber roughness and structure of the interface. They also suggested that optimizing the process parameters, surface modification, modification of printing head, and treatment is beneficial for improvement of the shear performance of 3D- printed composites [30].

10.2.7 Hardness Properties of 3D Printed Parts

Hardness is defined as the resistance of materials to localized plastic deformation. Hardness ranges from super hard materials such as diamonds to soft metals and down to plastics and soft tissues. Hardness can be assessed by a number of techniques including indentation, scratch, and rebound hardness measurements. There are many standards to measure hardness in different scales.

ASTM E384 is used to determine micro indentation hardness of materials (including plastics) based on the Knoop and Vickers hardness scales. This standard also includes specifications of machines' calibration and test blocks of both scales. ASTM D2240 is defined for the Shore Durometer Hardness test standard and covers the relative hardness of soft materials, usually elastomers. This test is also able to measure the depth of penetration of an indenter into the material under specified load and time. ASTM D785 is the test method for the Rockwell Hardness Test, which is more common for 3D-printed materials. This method measures the indention hardness of plastics by measuring the depth of indentation of an indenter in a material. Indenters are usually round hardened steel balls of specific diameters. ISO has similar standards for the Rockwell Hardness Test, namely, ISO 2039-1, wherein the method consists of forcing a hardened steel ball under a certain load onto the surface of a sample, and then the depth of indentation is measured, and ISO2039-2 wherein the Rockwell hardness number is derived from the net increase in depth of indentation as the load on an indenter is increased using major and minor loads [1].

Hanon et al. demonstrated that the hardness properties of 3D printed PLA parts fabricated by the FDM technique depend on build orientation, raster direction angle, and layer height. Hence, a stronger relationship was observed between hardness and print orientation [31]. Mishra et al. similarly observed the collective effect of infill pattern and infill density on the impact and hardness properties of 3D printed PLA [32]. Maguluri et al. explored the role of three printing parameters, including fill density, extrusion temperature, and printing speed on the hardness of PLA parts.

The study results have shown that extrusion temperature profoundly influences the hardness of the 3D-printed PLA specimens, while printing speeds have a much smaller impact on it [33].

10.2.8 Fatigue Properties of 3D Printed Parts

The fatigue limit refers to the stress level of less than an infinite number of loading cycles applied to a material without causing failure [1]. To evaluate the fatigue behavior of the material, Eq. (10.11) can be used to derive the parameters of fatigue properties:

$$\sigma_a = \sigma_f'(2N_f)^b \tag{10.11}$$

where, b is the fatigue strength exponent, σ_f' is the fatigue strength coefficient, σ_a is the stress amplitude, and N_f is the fatigue lifetime.

ASTM D7791 and ISO 13003 standards are recommended to test flexure fatigue of plastics. Both standards in terms of content are the same, but technically different. In both tests, the loading is sinusoidal. ASTM D7791 utilizes a three-point or four-point loading with cycling occurring in positive and negative directions, with either stress or strain mode vs cycle numbers. The R ratio is -1 and the stress or strains do not exceed the proportional limit. The test ends at either failure or reaching 10^7 cycles. ISO 13003 calculates the ultimate tensile/flexural strength for fatigue loading rate. In strain mode, the test finishes when the damage level reaches to specimen stiffness reduction of 20%. Four fatigue levels are tested according to the fatigue life of interest or the range of stress/strain of interest. Similarly, strain or stress vs. the number of cycles is reported. Both standards cannot cover 3D-printed materials with anisotropy. ASTM D6115 and ISO 15850 are also two standards which analyze fatigue delamination or crack propagation. Both of these standards are specifically applicable to the measurement of fracture energy in the interlaminar region of a fiber composite [2].

Letcher et al. investigated effects of three print directions of 0°, 45°, and 90° on the fatigue behavior of 3D-printed PLA parts and understood that the specimen with the 90° direction had the least resistance, while the specimen with the 45° direction offered the highest fatigue endurance limit [34]. Gomez-Gras et al. studied the influence of raster patterns, squared and hexagonal lattices, layer height, infill density, nozzle diameter, and the printing velocity, on the fatigue response of PLA samples The best fatigue results were obtained for 75% of the raster, the hexagonal raster pattern, the nozzle diameter of 0.5 mm, and the thickness of 0.3 mm [35]. Vidrih et al. explored the fatigue behavior of unidirectional 3D-printed continuous carbon fiber-reinforced polymer (CFRP) tension straps with a polyamide matrix (PA12), which showed that the fatigue endurance limit of the 3D-printed and post-consolidated CFRP strap is appropriate compared to steel tendons. However, it is still 20% lower than conventionally produced CFRP straps using out-of-autoclave unidirectional carbon fiber prepregs [36]. Terekhina et al. demonstrated that build orientation might impact fatigue behaviors of semi-crystalline polyamide 6 produced by FFF. Additionally, porosity can lead to a reduction in fatigue endurance [37].

10.2.9 Creep Properties of 3D Printed Parts

The creep induces permanent deformation to materials under specific load and is time- and temperature-dependent. At room temperature, creep deformation can be often neglected when applied stresses are significantly below the material's yield limit, but it is crucial for components requiring higher precisions. Creep becomes even more critical, as it is considered a life-limiting issue under higher loads and elevated temperatures. Understanding the creep and recovery behavior of polymer composites is vital, but not limited to their application in the aerospace and automotive industries [38].

The creep rate can be calculated as follows:

$$\dot{\varepsilon} = A\sigma^n \exp\left(\frac{-Q}{RT}\right) \tag{10.12}$$

where, Q is activation energy, R is gas constant, T is temperature, A is constant, and n is stress exponent.

Generally, creep standards recommend methods to measure dimensional changes of specimens under load as a function of different exposure environments such as temperature, aqueous, or surfactant solutions. The applied load can be different including tensile, compression, and flexural. ASTM D2990-09 is main standard for the creep testing of 3D-printed polymeric materials. This standard recommends ASTM D543 to specify solution composition for sample immersion. The ISO standard equivalent of ASTM D2990-09 is ISO 899. There are restrictions on the ratio of the length to the cross-section. The method recommends testing at a minimum of two different test temperatures within the use range of the material to understand the effect of temperature. Loading at seven stress levels to produce creep-rupture at different times up to 3000 h provides a measure of long-term performance. Design data for creep is obtained by testing materials at different stress levels to produce 1% strain in 1000 h. There is no guidance provided for anisotropic samples, such as fiber composites [2].

Rashid and Ko evaluated the mechanical properties of continuous fiber-reinforced 3D-printed (CFR3DP) composites at elevated temperatures.

The effect of different fiber reinforcements (Kevlar, fiberglass, and high-strength high-temperature fiberglass) and temperatures on the creep and recovery behavior of CFR3DP Onyx composites were examined both experimentally and numerically. According to the obtained results, a significant drop in maximum and residual strains was observed due to the introduction of fibers. However, the creep resistance of all the materials was affected at higher temperatures. Minimum creep strain was observed for Onyx-FG at 120 °C; however, at the same temperature, the minimum residual strain was obtained for Onyx-KF [38]. Niaza et al. studied the long-term creep behavior of PLA and polylactide/hydroxyapatite (PLA/HA) composites and reported positive effects of nanoparticles on hardness and creep resistance [39]. Salazar-Martin et al. tested the creep behavior of 3D-printed polycarbonate parts by FDM and concluded that the printing orientation significantly affects the elastic and viscoelastic components of the creep deformation. Similarly, Mohammadizadeh et al. found that the creep behavior of continuous fiber reinforced additively manufactured (CFRAM) components is a viscoelastic behavior with large deformation in the first stage, following a sharp decline in strain, and a gradual decrease in strain. Furthermore, increasing the temperature increases the strain percentage, and fiber reinforcement reduces the strain amount [40].

10.3 Conclusion

3D printing technologies have gained significant attention in fabricating different polymeric components over the last few years. Hence, the mechanical behaviors of such materials under different conditions play a key role in their applications. The most important mechanical properties addressed in this chapter were module, tensile, compressive, flexural, impact, shear, hardness, fatigue, and creep. According to mentioned studies in this regard, a large number of factors influenced these mechanical properties in polymetric 3D-printed parts. These factors were mainly a function of type and also the fabrication method of the polymetric component. Commonly used ASTM and ISO test standards were introduced in this regard.

References

1 Dizon, J.R.C., Espera, A.H., Chen, Q., and Advincula, R.C. (2018). Mechanical characterization of 3D-printed polymers. *Additive Manufacturing* 20: 44–67. https://doi.org/10.1016/J.ADDMA.2017.12.002.

2 Forster, A. (2015). *Materials Testing Standards for Additive Manufacturing of Polymer Materials: State of the Art and Standards Applicability*. Gaithersburg, MD: National Institute of Standards and Technology.

3 Arpan, M.F.Z.I.B.M. and Hoong, L.J. (2019). Effect of printing temperature on mechanical properties of copper metal polylactide acid. *AIP Conference Proceedings* 2137: 040003. https://doi.org/10.1063/1.5121001.

4 Vălean, C., Marşavina, L., Mărghitaşl, M. et al. (2020). Effect of manufacturing parameters on tensile properties of FDM printed specimens. *Procedia Structural Integrity* 26: 313–320. https://doi.org/10.1016/J.PROSTR.2020.06.040.

5 Dwiyati, S.T., Kholil, A., Riyadi, R., and Putra, S.E. (2019). Influence of layer thickness and 3D printing direction on tensile properties of ABS material. *Journal of Physics Conference Series* 1402: 066014. https://doi.org/10.1088/1742-6596/1402/6/066014.

6 Galeja, M., Hejna, A., Kosmela, P., and Kulawik, A. (2020). Static and dynamic mechanical properties of 3D printed ABS as a function of raster angle. *Maternité* 13: 297. https://doi.org/10.3390/MA13020297.

7 Wang, X., Zhao, L., Fuh, J.Y.H., and Lee, H.P. (2019). Effect of porosity on mechanical properties of 3D printed polymers: experiments and micromechanical modeling based on X-ray computed tomography analysis. *Polymer* 11: 1154. https://doi.org/10.3390/POLYM11071154.

8 Garzon-Hernandez, S., Garcia-Gonzalez, D., Jérusalem, A., and Arias, A. (2020). Design of FDM 3D printed polymers: an experimental-modelling methodology for the prediction of mechanical properties. *Materials and Design* 188: 108414. https://doi.org/10.1016/J.MATDES.2019.108414.

9 Subeshan B, Alonayni A, Rahman MM, Asmatulu E. Investigating compression strengths of 3D printed polymeric infill specimens of various geometries. *Conference Proceedings spie*, vol. 10597, SPIE; 2018, p. 89–94. https://doi.org/10.1117/12.2296651.

10 Nomani, J., Wilson, D., Paulino, M., and Mohammed, M.I. (2020). Effect of layer thickness and cross-section geometry on the tensile and compression properties of 3D printed ABS. *Materials Today Communications* 22: 100626. https://doi.org/10.1016/J.MTCOMM.2019.100626.

11 Wang, K., Cai, R., Zhang, Z. et al. (2021). Compressive behaviors of 3D printed polypropylene-based composites at low and high strain rates. *Polymer Testing* 103: 107321. https://doi.org/10.1016/J.POLYMERTESTING.2021.107321.

12 Valvez, S., Silva, A.P., and Reis, P.N.B. (2022). Compressive behaviour of 3D-printed PETG composites. *Aerospace* 9: 124. https://doi.org/10.3390/AEROSPACE9030124.

13 Krzikalla, D., Měsíček, J., Halama, R. et al. (2022). On flexural properties of additive manufactured composites: experimental, and numerical study. *Composites Science and Technology* 218: 109182. https://doi.org/10.1016/J.COMPSCITECH.2021.109182.

14 Chen, J., Hao, N., Zhao, T., and Song, Y. (2021). Flexural properties and failure mechanism of 3D-printed grid beetle elytron plates. *International Journal of Mechanical Sciences* 210: 106737. https://doi.org/10.1016/J.IJMECSCI.2021.106737.

15 Ibrahim, Y., Melenka, G.W., and Kempers, R. (2019). Flexural properties of three-dimensional printed continuous wire polymer composites. *Materials Science and Technology* 35: 1471–1482. https://doi.org/10.1080/02670836.2019.1630085.

16 Araya-Calvo, M., López-Gómez, I., Chamberlain-Simon, N. et al. (2018). Evaluation of compressive and flexural properties of continuous fiber fabrication additive manufacturing technology. *Additive Manufacturing* 22: 157–164. https://doi.org/10.1016/J.ADDMA.2018.05.007.

17 Alshamrani, A.A., Raju, R., and Ellakwa, A. (2022). Effect of printing layer thickness and postprinting conditions on the flexural strength and hardness of a 3D-printed resin. *BioMed Research International* 2022: 1–9. https://doi.org/10.1155/2022/8353137.

18 Li, Y. and Lou, Y. (2020). Tensile and bending strength improvements in PEEK parts using fused deposition modelling 3D printing considering multi-factor coupling. *Polymer* 12: 2497. https://doi.org/10.3390/POLYM12112497.

19 Patterson, A.E., Pereira, T.R., Allison, J.T., and Messimer, S.L. (2019). IZOD impact properties of full-density fused deposition modeling polymer materials with respect to raster angle and print orientation. *Proceedings of the Institution of Mechanical Engineers, Part C: Journal of Mechanical Engineering Science* 235: 1891–1908. https://doi.org/10.1177/0954406219840385.

20 Lay, M., Thajudin, N.L.N., Hamid, Z.A.A. et al. (2019). Comparison of physical and mechanical properties of PLA, ABS and nylon 6 fabricated using fused deposition modeling and injection molding. *Composites. Part B, Engineering* 176: 107341. https://doi.org/10.1016/J.COMPOSITESB.2019.107341.

21 Upadhyay, K., Dwivedi, R., and Singh, A.K. (2016). Determination and comparison of the anisotropic strengths of fused deposition modeling

P400 ABS. In: *Advances in 3D Printing & Additive Manufacturing Technologies*, 9–28. https://doi.org/10.1007/978-981-10-0812-2_2/COVER.

22 Wang, L., Gramlich, W.M., and Gardner, D.J. (2017). Improving the impact strength of Poly(lactic acid) (PLA) in fused layer modeling (FLM). *Polymer (Guildf)* 114: 242–248. https://doi.org/10.1016/J.POLYMER.2017.03.011.

23 Kabir, S.M.F., Mathur, K., and Seyam, A.F.M. (2020). Impact resistance and failure mechanism of 3D printed continuous fiber-reinforced cellular composites. *Journal of the Textile Institute* 112: 752–766. https://doi.org/10.1080/00405000.2020.1778223.

24 Tsou, C.H., Yao, W.H., Wu, C.S. et al. (2019). Preparation and characterization of renewable composites from Polylactide and Rice husk for 3D printing applications. *Journal of Polymer Research* 26: 1–10. https://doi.org/10.1007/S10965-019-1882-6/METRICS.

25 Peng, F., Zhao, Z., Xia, X. et al. (2018). Enhanced impact resistance of three-dimensional-printed parts with structured filaments. *ACS Applied Materials & Interfaces* 10: 16087–16094. https://doi.org/10.1021/ACSAMI.8B00866/SUPPL_FILE/AM8B00866_SI_001.PDF.

26 Tezel, T., Ozenc, M., and Kovan, V. (2021). Impact properties of 3D-printed engineering polymers. *Materials Today Communications* 26: 102161. https://doi.org/10.1016/J.MTCOMM.2021.102161.

27 Cantrell, J., Rohde, S., Damiani, D. et al. (2017). Experimental characterization of the mechanical properties of 3D printed ABS and polycarbonate parts. *Conference Proceedings of the Society for Experimental Mechanics Series* 3: 89–105. https://doi.org/10.1007/978-3-319-41600-7_11/COVER.

28 Elmrabet, N. and Siegkas, P. (2020). Dimensional considerations on the mechanical properties of 3D printed polymer parts. *Polymer Testing* 90: 106656. https://doi.org/10.1016/J.POLYMERTESTING.2020.106656.

29 Caminero, M.A., Chacón, J.M., García-Moreno, I., and Reverte, J.M. (2018). Interlaminar bonding performance of 3D printed continuous fibre reinforced thermoplastic composites using fused deposition modelling. *Polymer Testing* 68: 415–423. https://doi.org/10.1016/J.POLYMERTESTING.2018.04.038.

30 Yang, C., Tian, X., Liu, T. et al. (2017). 3D printing for continuous fiber reinforced thermoplastic composites: mechanism and performance. *Rapid Prototyping Journal* 23: 209–215. https://doi.org/10.1108/RPJ-08-2015-0098.

31 Hanon, M.M., Dobos, J., and Zsidai, L. (2021). The influence of 3D printing process parameters on the mechanical performance of PLA polymer and its correlation with hardness. *Procedia Manufacturing* 54: 244–249. https://doi.org/10.1016/J.PROMFG.2021.07.038.

32 Mishra, P.K., Senthil, P., Adarsh, S., and Anoop, M.S. (2021). An investigation to study the combined effect of different infill pattern and infill density on the impact strength of 3D printed polylactic acid parts. *Composites Communications* 24: 100605. https://doi.org/10.1016/J.COCO.2020.100605.

33 Maguluri, N., Suresh, G., and Guntur, S.R. (2022). Effect of printing parameters on the hardness of 3D printed poly-lactic acid parts using DOE approach. *IOP Conference Series: Materials Science and Engineering* 1248: 012004. https://doi.org/10.1088/1757-899X/1248/1/012004.

34 Letcher, T. and Waytashek, M. (2015). Material property testing of 3D-printed specimen in PLA on an entry-level 3D printer. *Advanced Manufacturing* 2A: 1–8. https://doi.org/10.1115/IMECE2014-39379.

35 Gomez-Gras, G., Jerez-Mesa, R., Travieso-Rodriguez, J.A., and Lluma-Fuentes, J. (2018). Fatigue performance of fused filament fabrication PLA specimens. *Materials and Design* 140: 278–285. https://doi.org/10.1016/J.MATDES.2017.11.072.

36 Vidrih, T., Winiger, P., Triantafyllidis, Z. et al. (2022). Investigations on the fatigue behaviour of 3D-printed continuous carbon fibre-reinforced polymer tension straps. *Polymer* 14: 4258. https://doi.org/10.3390/POLYM14204258.

37 Terekhina, S., Tarasova, T., Egorov, S. et al. (2020). The effect of build orientation on both flexural quasi-static and fatigue behaviours of filament deposited PA6 polymer. *International Journal of Fatigue* 140: 105825. https://doi.org/10.1016/J.IJFATIGUE.2020.105825.

38 Al Rashid, A. and Koç, M. (2021). Creep and recovery behavior of continuous fiber-reinforced 3DP composites. *Polymer* 13: 1644. https://doi.org/10.3390/POLYM13101644.

39 Niaza, K.V., Senatov, F.S., Stepashkin, A. et al. (2017). Long-term creep and impact strength of biocompatible 3D-printed PLA-based scaffolds. *Nano Hybrids Compos* 13: 15–20. https://doi.org/10.4028/WWW.SCIENTIFIC.NET/NHC.13.15.

40 Seifans, A.M., Ayyagari, S., and Al-Haik, M. (2021). Elastic/viscoplastic characterization of additively manufactured composite based on continuous carbon fibers. *Aerospace Science and Technology* 111: 106562. https://doi.org/10.1016/J.AST.2021.106562.

11

Thermal Modeling of Material Extrusion Additive Manufacturing (MEX)

José A. Covas[1], *Sidonie F. Costa*[2], *and Fernando M. Duarte*[1]

[1] *Institute for Polymers and Composites (IPC), University of Minho, Guimarães, Portugal*
[2] *Centre for Research and Innovation in Business Sciences and Information Systems, School of Management and Technology (ESTGF), Porto Polytechnic Institute (IPP), Felgueiras, Portugal*

11.1 Introduction

In material extrusion additive manufacturing (MEX), a thermoplastic polymer system (generally, in the shape of a circular filament or as pellets or powder) is compressed, molten, and extruded as a filament thread through a nozzle, while the latter, or the build platform, moves following a predefined printing path. 3D structures are built as vertical series of horizontal cross-sections, each consisting of threads(s) deposited in a prespecified manner. Geometrically complex parts can thus be manufactured in this way. Bonding between adjacent threads is necessary for the physical cohesiveness of the part. Indeed, the mechanical properties, surface roughness, dimensional tolerances, and residual stresses of a 3D-printed part are strongly influenced by bonding and by the morphology developed upon cooling. Effective bonding requires the extensive diffusion and re-entanglement of the polymer macromolecules across the interfaces of adjacent filament threads, which depend on the rheological properties of the polymer, the local temperatures, and associated times.

Consequently, for a given geometry of the part to be printed and for a specific set of processing conditions, MEX involves heating and melting of a material, melt pressure flow through a nozzle, deposition of the extruded thread first onto a build platform and subsequently onto previously deposited layer(s), and cooling/solidification.

This chapter focuses on modeling the thermal aspects of the deposition and cooling stages of MEX, with the aim of better understanding the correlations between the printing parameters, the resulting temperature profiles

Industrial Strategies and Solutions for 3D Printing: Applications and Optimization, First Edition. Edited by Hamid Reza Vanaei, Sofiane Khelladi, and Abbas Tcharkhtchi.

and the extent of bonding. Thermal modeling could also be used to support part design, setting the corresponding printing parameters, and for process optimization purposes.

11.2 Thermal Modeling of MEX

Modeling the entire MEX processing sequence, i.e. from forcing the solid filament into the liquefier to cooling the part, is extremely complex, as it entails a wide range of physical and thermal phenomena, including heat transfer, molecular diffusion, phase change, and morphology development, as well as non-isothermal viscoelastic shear and extensional flows [1]. For example, Xia et al. [2, 3] proposed full 3D computational fluid dynamic (CFD) resolved simulations, including heat transfer, shear thinning behavior, solidification, and residual stresses. Despite the high accuracy of this approach, it is not only computationally quite demanding but also requires full, complex, and lengthy material characterization.

In practice, it has been demonstrated that knowing the temperature evolution of the thread(s) during the deposition stage is decisive for understanding and predicting part quality under service. In fact, the existing heating and subsequent cooling processes can create high thermal gradients due to the low polymer thermal conductivity, inducing residual stresses that can induce warpage and distortions on the printed part. Thermal gradients can also affect the mechanical performance, namely, yield strength and ductility [4].

Efforts to predict the temperature history during deposition/cooling of the threads began in the last decade of the last century and continue today due to its complexity. Yardimci and Güceri [5] were some of the first to underline the importance of a heat transfer analysis in MEX. They proposed a 1D formulation considering 1D filament threads modeled as grid blocks. Interactions with the environment and adjacent filament threads were included. Convection boundary conditions were added to the end surfaces of the threads. Based on different ambient cooling conditions, the bonding potential of adjacent threads was estimated. Later, Thomas and Rodriguez [6] and Rodriguez et al. [7] developed an analytical 2D heat transfer model assuming rectangular cross sections for the extruded threads in a vertical stack, neglecting all contact resistances. Instead, free convection boundary conditions were imposed on the outer surfaces. Bonding was predicted using a wetting-diffusion model based on the reptation theory, and it was shown that lower cooling rates promote stronger bonding.

In recent years, several authors adopted different approaches and degrees of sophistication to compute the temperature evolution during cooling in

MEX, the problem being typically solved, as illustrated in Figure 11.1. It is usually assumed (Figure 11.1a) that at each time interval Δt of the total deposition time a Δx length increment of thread is deposited. Three different thermal states can then be defined:

(i) the Δx increment is isothermal;
(ii) the temperature of the Δx increment varies radially at each cross-section but is constant along its length (2D approach); and
(iii) the temperature of the Δx increment varies both radially and lengthwise (3D approach).

The gradual deposition of the thread/part is thus described by an increasing series of contiguous Δx elementary lengths, each associated with a time interval Δt. The heat exchanges for each Δx, i.e. the local thermal boundary conditions, must be defined and taken into consideration when solving the energy equation. They depend on the geometry of the part, build orientation, deposition strategy, and whether it is the first layer or any other layer being printed. As illustrated in Figure 11.1b, the boundary conditions can comprise:

(i) contact (conduction) between filament thread and build platform;
(ii) contact (conduction) between adjacent threads;
(iii) natural and/or forced convention between thread and environment (e.g. build chamber); and
(iv) radiation between thread and the environment.

The computational strategy to track the material deposition involves the discretization of the entire build volume and the identification of the relevant intra-layer and inter-layer interactions. As displayed in Figure 11.1c, 1D (a vertical stack of filament threads) or 3D structures are typically computed.

The temperature evolution of each Δx thread is computed either analytically or numerically. In this case, finite differences, finite elements, or finite volume methods are adopted, often using commercial software packages such as ABAQUS®, MATLAB®, ANSYS®, or COMSOL®.

Several major contributions to the thermal modeling of MEX reported in the open literature are identified in Table 11.1, together with a short characterization of the strategy developed in terms of the modeling approach, solver utilized, thermal state of the length increment of thread Δx, boundary conditions for that increment, and computational strategy. Whenever relevant, computations performed based on the availability of the temperature evolution, such as on the extent of bonding, mechanical performance, warpage, or surface quality, are also included.

Costa et al. [8] proposed an analytical solution for the transient heat transfer in 3D MEX by sectioning the part into axial increments that are

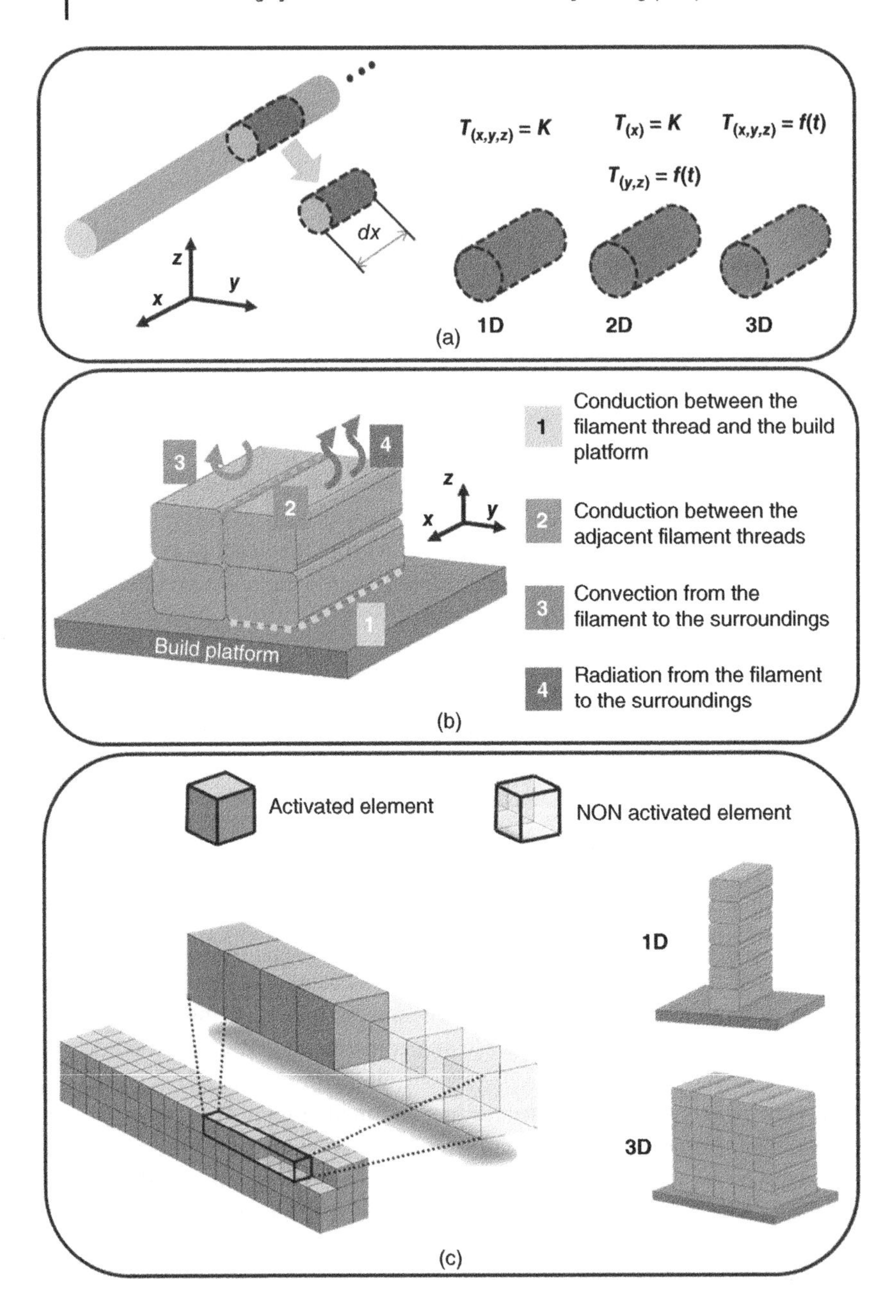

Figure 11.1 Main approaches usually adopted to model cooling in MEX. (a) thermal states of a length increment of thread Δx; (b) possible boundary conditions for increment Δx; and (c) – computational simulation approaches.

Table 11.1 Simulation studies reported in the literature (sorted by publication year) of the cooling of filament threads in MEX (see Figure 11.1 to identify A, B, and C).

Refs.	Modeling approach	Solver	A	B	C	Obs.
[5]	Finite volume method		1D	2,3	3D	Bonding potential distribution
[7]	Analytical model	ABAQUS®	2D	3	1D	Bonding
[8]	Analytical model	MATLAB®	1D	1,2,3	3D	Bonding
[9]	Analytical model	ANSYS®	1D	1,2,3	3D	Thermal stress distributions
[10]	Finite element model	ANSYS®	1D	3	3D	Multi-material interfacial bonding
[11]	Finite difference method		2D	1,2,3,4	3D	Temperature
[12]	Finite element model	COMSOL®	1D	1,2,3	1D	Temperature
[13]	Finite element model	ABAQUS®	2D	1,2,3,4	3D	Temperature
[14]	Analytical model		1D	1,2,3	1D	Mechanical properties, filling patterns, and raster angles
[15]	Analytical model		1D	1,2,3	3D	Surface roughness
[16]	Finite element model	COMSOL®	2D	1,2,3	1D	Degree of healing
[17]	Finite difference method	MATLAB®	2D	1,2,3	1D	Bonding
[18]	Finite difference method		2D	1,2,3	3D	Bonding, tear resistance, and warping
[19]	Finite difference method	MATLAB®	1D	1,2,3	3D	Tensile load at failure for complex geometries and infill densities
[20]	Finite element model	C++ library	1D	1	3D	Temperature
[21]	Finite difference method	LabVIEW®	1D	1,3	3D	Bonding, mechanical performance
[22]	Finite element model	ANSYS®	1D	1,2,3		Accuracy and efficiency of heat transfer simulations of FFF printing
[23]	Finite element model	ABAQUS®	2D	1,2,3,4	3D	Part distortion

added at discrete times according to the predefined printing path, and assuming that each one has uniform temperature across the transversal area and length. The temperature evolution was computed with a code developed in MATLAB®, considering the most relevant boundary conditions, such as convection with the environment, conduction between adjacent layers and with the build platform, as well as the contact area between threads. Moreover, the temperature evolution was correlated with interlayer adhesion. The predictions were validated experimentally.

Concurrently, Zhou et al. [9] performed a thermal analysis of a single 3D filament thread on a built platform using the ANSYS® software. The latent heat of phase change was included, as well as the temperature-dependent heat conduction and heat capacity. The printing process was modeled by a stepwise activation of the bead, and temperature profiles were shown at different times. Both convective and radiative heat transfer phenomena were considered to develop a 3D analytical model.

In the following year, Yin et al. [10] reported the use of the finite element method to predict the interfacial temperature history. A 3D transient heat transfer analysis was performed numerically on a multi-material MEX process with the governing equation proposed by Rodriguez et al. [7]. The temperature at the bottom surface of the part was set equal to the temperature of the build platform and heat losses by convection were included. The problem was solved using the commercial software ANSYS®, by means of the element birth/death feature to mimic the continuous filament deposition.

Zhang and Shapiro [11] computed the temperature field in a MEX-printed part using a finite-differences method. Multiple contacts between printed threads were taken into account, but the thermal computations were only performed on increments directly influenced by the deposition of a new thread. D'Amico and Peterson [12] modeled the cooling of vertical filament stacks. The approach utilized the commercial software COMSOL®, using element death and birth effects to represent the addition of polymer during the deposition sequence. A good agreement with experimental data was reported. Roy et al. [13] solved the energy equation together with the boundary and initial conditions using ABAQUS®. The geometry change during material deposition was also accounted for through the element birth and death simulation scheme in finite elements. The part geometry was discretized using a self-developed code.

As seen from the above examples, the equations governing heat transfer have been solved using one of three approaches. Considering the Δx thread increment as isothermal allows the usage of analytical models and, therefore, the reduction of the computation time. However, this simplification is only valid for Biot numbers lower than one, i.e. for the so-called thermally simple problems involving uniform temperature fields inside the body under study [18]. In order to relax this simplification, finite differences [e.g. 11, 17, 18, and 21] and finite elements [e.g. 10, 12, 13, 16, 20, 22, and 23] have been used by many authors.

When several thousands of filament thread elements must be considered to define the geometry to be modeled, and it is also necessary to define small time increments to ensure accuracy and convergence of the results, it is convenient to adopt a computational strategy minimizing the computation time. This is often done by adopting "birth and death" elements. "Element death" refers to the case when the propriety, such as stiffness or conductivity, of an element is reduced to a very low level, $1E^{-6}$, by default. In other words, the properties of those "dead" elements would barely influence the surrounding elements, and so the computations for those elements can be suspended. In the same manner, if an element is "born" or reactivated, the property will be restored to the normal value [9, 11].

As far as the boundary conditions are concerned, most authors take into account the heat conduction between threads and build platform, conduction between neighboring threads, and heat losses to the environment by natural or forced convection. A few authors have also included radiation from the nozzle in their calculations [e.g. 23]. However, Trofimov et al. [23] found that radiation heat transfer from the nozzle had a negligible influence on the thermal fields of the part due to the high speed of MEX, convection and conduction being the dominant thermal mechanisms. The same conclusion was reported by Costa et al. [24].

Defining the correct values for the heat transfer coefficients to be used in the computations is a difficult but crucial task, as it strongly influences the accuracy of the predictions. Quite distinct values can be extracted from the literature for the same situation. For example, considering the heat losses to the environment by natural convection, the convective heat transfer coefficient has been set to 62, 88, 30, 29, and 19 W/m^2 °C in Refs. [8, 9, 16, 18, 19], respectively. Recently, Ramos et al. [22] studied the effect of the value of this coefficient in achieving a more accurate thermal boundary condition expressing the natural convection between the printed material and the environment. It was shown that the temperature fields could be simulated with a high level of accuracy using values between 20 and 30 W/m^2 °C. The shape of the thread cross-section, the existence of voids, and the infill percentage are also relevant parameters influencing heat transfer by conduction since they affect the contact area between adjacent threads and between the latter and the build platform. Although for computational purposes most authors assume the thread cross-section as a perfect square (with full contact with contiguous threads) [e.g. 8, 13, and 15], some take into account the actual contact area. For example, Costa et al. [8], use as input data the percentage surface perimeter/area in contact with neighboring filament threads, build platform, and environment. Ramos et al. [22] studied the effect of discretization simplifications on the accuracy of modeling structures with varying void sizes, infill densities, and infill patterns, together with experimental thermal measurements. They concluded that accurate heat transfer simulation in such structures is possible as long as the infill density is respected, whereas the presence of voids can be disregarded.

11.3 A Thermal Model for Heat Transfer and Bonding

The model presented here assumes that the Δx thread increment is isothermal, and thus it solves analytically the energy equation but takes in the analysis of 3D structures and considers the local boundary conditions [e.g. 8, 11–13, and 17–20].

The importance of the physical contacts arising during deposition on the cooling of the threads is illustrated in Figure 11.2, which considers a simple parallelepiped-shaped 3D structure composed of three vertical layers (z-axis) of four filament threads (x–y plane, each with a diameter of 0.3 mm and a length $L = 70$ mm). Threads are deposited in the sequence 1–9, as identified in Figure 11.2a, each deposition taking 2.8 s. Figure 11.2b displays the predicted temperature evolution of all the threads at a vertical plane y–z approximately at their mid-length. T_E represents the temperature of the build chamber (70 °C), while T_g represents the glass transition temperature of the polymer (in this example, an acrylonitrile butadiene styrene [ABS]). If only thread 1 was deposited, its temperature would gradually decrease from the extrusion temperature (270 °C) toward T_E. However, the Figure shows the existence of various reheating peaks during the cooling of this thread. They result from the combination of two contributions to heat transfer:

(a) The progressive contact with new and hotter threads (in this case, threads 2 and 8);
(b) The fact that the thermal effect of these new contacts spreads throughout the build volume. Indeed, the Figure shows that the cooling of thread 1 is affected by the deposition of the remaining 11 threads, although only threads 2 and 8 are in direct physical contact with it.

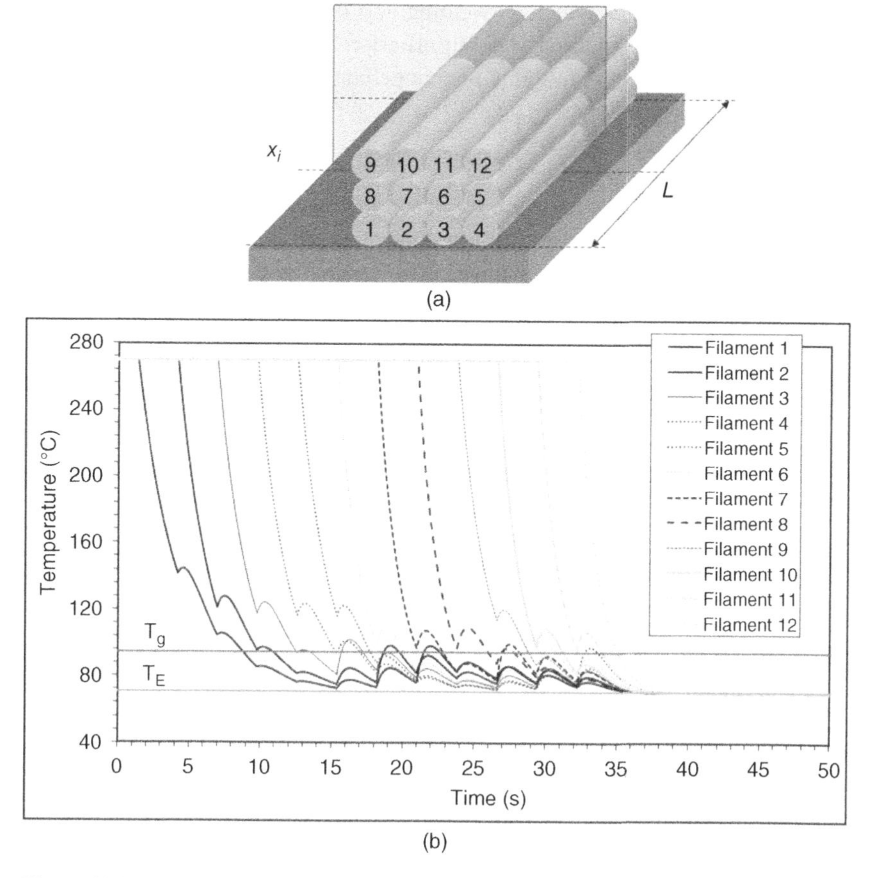

Figure 11.2 (a) Part to build and sequence of deposition; (b) Temperature evolution with time of all nine threads at the vertical plane identified in (a). An ABS thread with a diameter of 0.3 mm is extruded at 270°C, at the speed of 0.025 m/s, inside a build chamber kept at 70°C (Reproduced from Ref. [25]).

In the example depicted in Figure 11.2, the reheating peaks can reach as much as 24°C relative to the base cooling curves. In a few cases, threads that had cooled below T_g are reheated above this temperature when contacted by a new thread (e.g. threads 2 and 7, and 5 and 12, belonging to different layers), which is mandatory for proper bonding.

Indeed, bonding between two contiguous threads will only take place satisfactorily if the two interfaces contact each other at a temperature above Tg during a period that will depend on that same temperature. The higher the temperature, the lower the required time for bonding. This "healing process" has been studied by a number of authors, as it applies to other phenomena such as welding and droplet coalescence. Yang and Pitchumani [26] proposed a healing criterion that is based on a fundamental formulation of the reptation of polymer chains:

$$D_h(t) = \left[\int_0^t \frac{1}{t_w(T)} dt \right]^{1/4}, \tag{11.1}$$

where $t_w(T)$ is the temperature-dependent welding time, i.e. the time required to achieve the maximum bond strength. Bonding develops when $Dh(t) \geq 1$. $t_w(T)$ must be obtained experimentally, as it is polymer dependent. This criterion has been adopted by a number of authors studying the thermal phenomena developing in MEX.

The temperature–time curves shown in Figure 11.2 were obtained using a MATLAB® code that solves the energy equation resulting from an energy balance (see below) by taking into account the heat transfer modes developing during the gradual deposition of the thread. This is done by triggering the applicable thermal boundaries at each time step along the deposition stage. These depend on the geometry of the part, build orientation (i.e. the rotation of the part in the manufacturing space around the axes of the machine coordinate system), fill pattern (that is, the toolpath pattern of the thread during deposition), and infill degree (from fully hollow to fully dense). Therefore, as schematized in Figure 11.3, which presents the general flowchart of the computer code, conduction with the build platform of the 3D printer or with contacting thread(s) (in the lower or same layer), or convection (natural or forced when a build chamber exists) may be locally considered. In the analysis, radiation is not taken into consideration due to its small effect [24].

In MEX, an energy balance performed on an elementary length Δx of thread being deposited yields:

Energy in at one face – Heat transfer by convection between the lateral surface(s) and the environment – Heat transfer by conduction between the lateral surface(s) and adjacent thread(s) or the build platform = Change in internal energy + Energy out at opposite face.

Given the low thermal conductivity of polymers and the small thread diameter (typically lower than 0.5 mm), axial heat conduction can be neglected. Thus, the energy equation becomes [8]:

$$\frac{\partial T_r(x,t)}{\partial t} = -\frac{P}{\rho CA}\left(h_{conv}\left(1 - \sum_{i=1}^{n} a_{r_i}\lambda_i\right)(T_r(x,t) - T_E) + \sum_{i=1}^{n} h_i a_{r_i}\lambda_i (T_r(x,t) - T_{r_i}) \right) \tag{11.2}$$

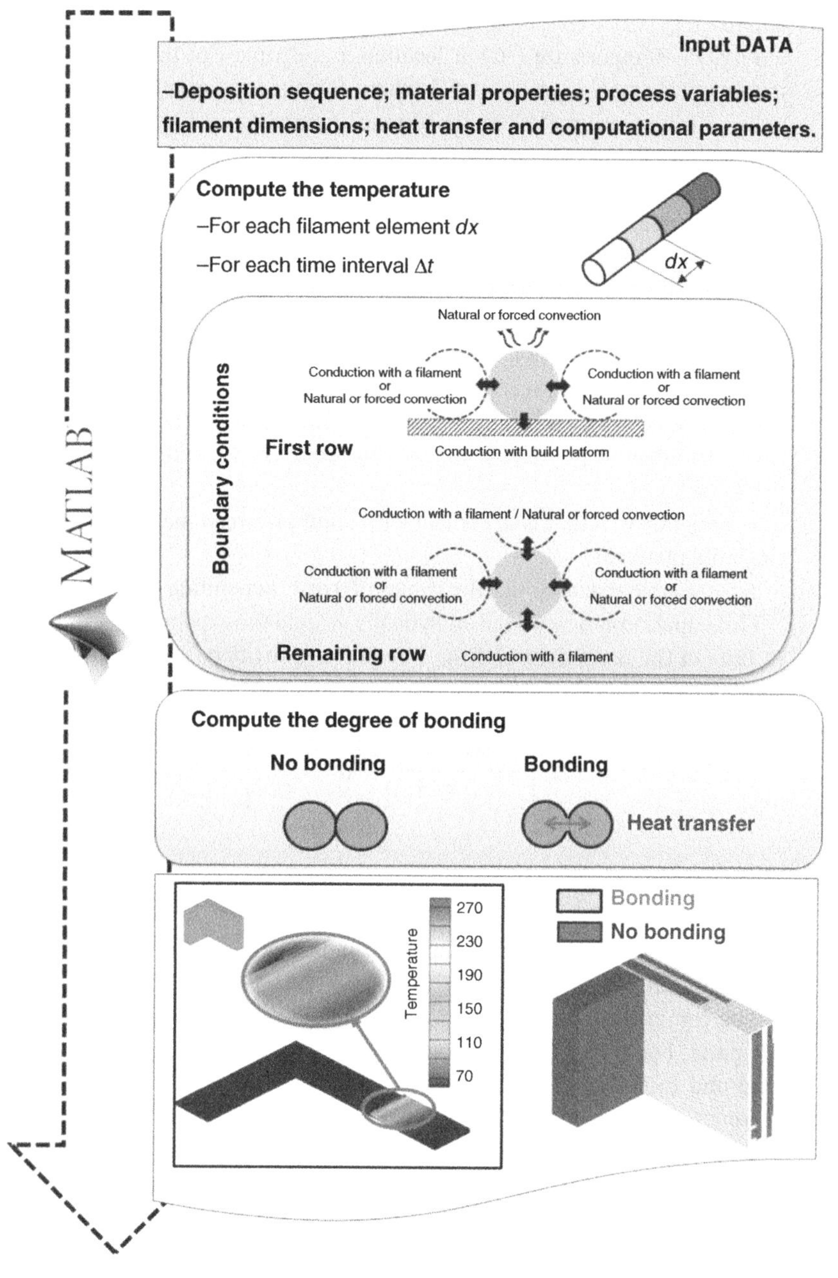

Figure 11.3 General flowchart of the computer code (Reproduced from Ref. [25]).

where:

$T_r(x,t)$ – Temperature (°C) at location x and time t of the rth filament increment (($r \in \{1, \ldots, N\}\{1, \ldots, N\}$, where N is the total number of threads

Tr_i – Temperature (°C) of the thread increment or build platform in contact with the rth filament's increment i ($r_i \in \{1, \ldots, N+1\}$, $r_i \neq r$).

T_E – Environment temperature (°C)

k – Thermal conductivity (W/m °C)

ρ – Specific weight (kg/m^3)

C – Specific heat capacity (J/kg °C)

P – Thread perimeter (m)

A – Area of thread cross-section (m^2)

h_{conv} – Convective heat transfer coefficient (W/m^2 °C)

h_i – Heat transfer coefficient for contact $i \in \{1, \ldots, n\}$ (W/m^2 °C)

n – Number of physical contacts with adjacent threads or with build platform

λ_i – fraction of P that is in contact with another thread increment or with the build platform

a_{ri} – variable that assumes the values of 0 or 1, depending on contacts

This equation can be solved analytically to yield the exponential evolution in time of the temperature of the thread along the deposition and cooling stage, which takes the form:

$$T_r(x,t) = C_1 \exp\left[\frac{-Pb(a_{r_1}, \ldots, a_{r_n})}{\rho CA}(t - t_r(x))\right] + Q(a_{r_1}, \ldots, a_{r_n}). \tag{11.3}$$

where $t_r(x)$ is the time at which an elementary length Δx of the rth thread is deposited and, consequently, begins to cool down or contact an adjacent thread(s) or the build platform. C_1, $b(a_{ri})$ and $Q(a_{ri})$ are expressions presented in Ref. [8].

The computer code developed is able to analyze geometrically complex 3D parts. For this purpose, a volume enveloping the part is initially created and then discretized into elementary parallelepiped volumes. These volumes are assigned with the value of 1 or 0, depending on whether they consist of polymer or not (i.e. air), respectively. A 3D matrix M_{mxnxp} is then obtained, where m is the number of layers, n is the maximum number of threads in one layer, and p is the maximum number of elementary volumes along the length of the threads. The size of the length increment Δx and the duration of the time interval Δt are related to processing parameters, such as nozzle diameter and printing velocity. During material deposition, the temperature at each time increment is calculated for all the existing

Δx increments until the process is completed. The code can take into account two materials (for example, the polymer material for making the part and a second material to function as support), which can be either be amorphous or partially crystalline. The nature of the polymer is important since modeling the cooling of a partially crystalline polymer must include the effect of the phase change from a liquid to a solid, with the consequent contribution of latent heat of solidification (a detailed analysis of the cooling of a partially crystalline thread can be found in Ref. [27]).

Upon coupling the healing criterion in Eq. (1) with the temperature–time computations, it becomes also possible to predict the degree of bonding throughout the build volume, which is an important measure of its quality. The general flowchart of the computer code is illustrated in Figure 11.3. The predictions obtained with the model both in terms of temperature and extent of bonding were shown to be generally in good agreement with carefully obtained experimental data [8, 28]. Therefore, thermal modeling of MEX can be successfully used to set the operating conditions that assure the manufacture/printing of a good quality part.

For instance, Figure 11.4 shows the effect of build orientation, extrusion temperature, and environment temperature on the bonding quality of a printed part. The part consists essentially of two orthogonal walls with distinct lengths but uniform thickness, creating an L-type geometry (Figure 11.4a). As seen in the insets of Figure 11.4b or Figure 11.4c, the part can be printed using six different build orientations, labeled A–F). Orientations E and F involve the use of a second polymer to build a support structure. In all cases, 100% infill, unidirectional, and aligned deposition sequences were selected (see Figure 11.4a). The typical material properties and reference process parameters for ABS were used in the computations (see details in Ref. [25]). Figure 11.4b shows that when extruding at 265°C or above, the part exhibits good quality regardless of the build orientation adopted. This is not surprising since most commercial 3D printers automatically set to 270°C the extrusion temperature for ABS. Bonding problems appear and increase with decreasing extrusion temperature. This is obvious since with decreasing extrusion temperatures the contacting threads have less time above T_g to properly bond. Also, orientations A and B have poorer performance since they necessitate higher extrusion temperatures (approximately 5–8°C) to reach the same quality as the remaining. The reasons for these differences are associated to the inherent characteristics of the deposition for each build orientation. As schematized in Figure 11.4a, in each layer (x–y plane), orientations A and B involve the deposition of long threads in one tab of the part and of short threads in the other tab. Differently, in the remaining build orientations, all threads have equal

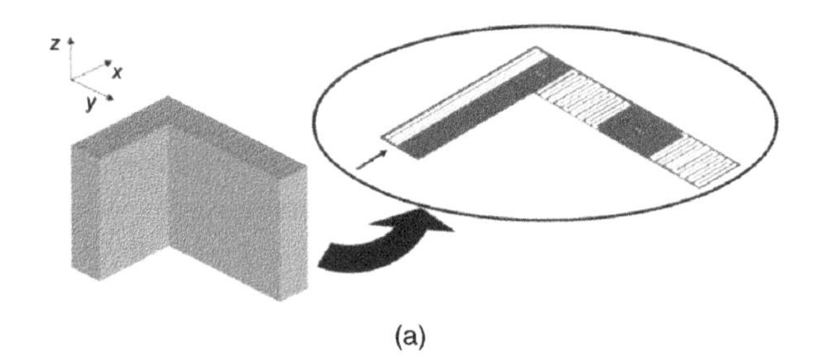

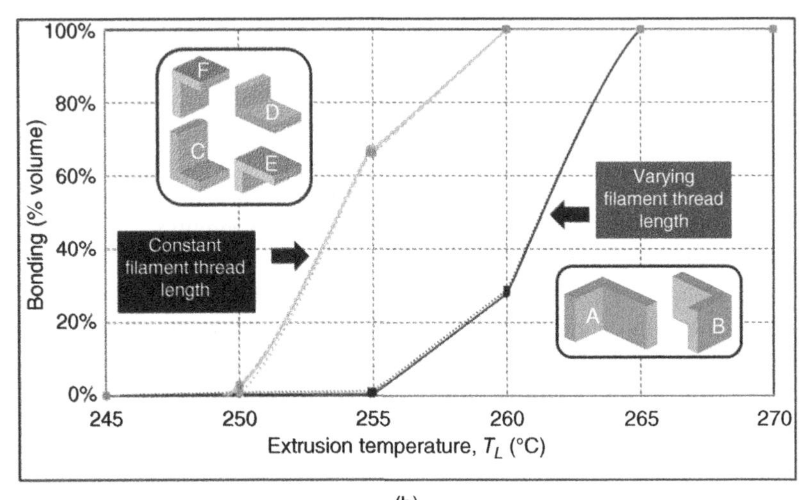

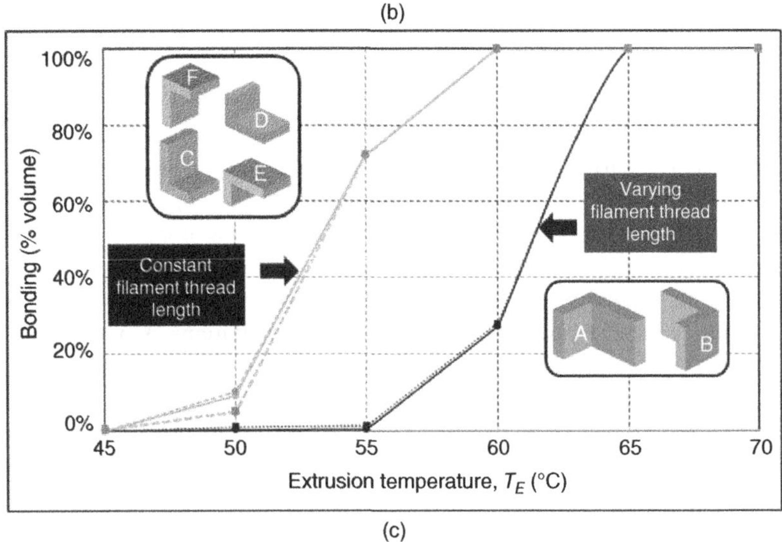

Figure 11.4 (a) Part to build and corresponding deposition sequence; (b) effect of build orientation and extrusion temperature on the bonding quality of the printed part; and (c) effect of build orientation and environment temperature on the bonding quality of the printed part (adapted from Ref. [25]).

lengths (with a value intermediate to those of the long and short threads of orientations A and B). At a given extrusion velocity, contacts between long threads arise at higher time intervals. This means that each of these filaments is likely to cool down significantly before contacting with a newly deposited thread. Consequently, reheating above T_g may be difficult or persist during a period too short to assure good bonding. Orientations C–F reveal a similar thermal behavior. Figure 11.4c demonstrates the importance of the environment temperature to bonding. As expected, the availability of a build chamber with controlled temperature may contribute positively to the quality of the final part. Obviously, the role of build orientation is not affected by this parameter.

In practice, orientation C is probably the most appropriate since it avoids the use of support material (with the corresponding much longer printing time) and minimizes the physical contact with the build platform. The latter may contribute to faster cooling if not set to a temperature slightly above the T_g of the polymer being printed, and it also increases the need of post-processing operations.

11.4 Printing a Tensile Test Specimen

This section demonstrates the effect of several printing parameters on the quality (in terms of bonding between contiguous threads) of a realistic 3D-printed part as predicted by the thermal model presented in the previous section. A tensile dumbbell test specimen (type 1BA, in accordance with the EN ISO527-2 standard), with a total length of 75 mm and a 5 × 2 mm cross-section in the narrower region, was chosen as case study. The part will be printed using ABS POLYLAC® PA-709S (Chi Mei Corporation, Tainan City, Taiwan), an extrusion grade with a melt flow index, MFI (220°C × 10 kg) = 4 ml/10 min, thermal conductivity, $k = 0.2\,\text{W/m}\,°\text{C}$, and specific heat, $C = 1800\,\text{J/kg}\,°\text{C}$. Table 11.2 presents the reference values selected for the main processing parameters as well as values of the heat transfer coefficients assumed in the calculations.

Figure 11.5 illustrates the effect of lowering the extrusion temperature from 270°C (which, as referred above, is the practical temperature usually adopted for printing ABS) down to 240°C. As expected, while at 270°C, the part should globally present good bonding between most threads, a decrease in the extrusion temperature seriously compromises the resulting quality. If the filament is extruded at 240°C, 96.1% of the volume of the part will exhibit insufficient bonding, whereas this value decreases to 0.5% at 270°C. Interestingly, even at this higher temperature, the model predicts bonding

Table 11.2 Main MEX processing parameters and heat transfer coefficients.

Property	Value
Deposition strategy	
Extrusion velocity, v (mm/s)	40
Extrusion temperature, T_L (°C)	270
Environment temperature, T_E (°C)	30
Build platform temperature, T_{sup} (°C)	100
Infill density (%)	100
Convective heat transfer coefficient, h_{conv} (W/m^2 °C)	56[a)]
Thermal contact conductance between adjacent filaments, h_i (W/m^2 °C)	h_i ε ∈ [10^{-4}; 220]
Thermal contact conductance between filament and support, h_{sup} (W/m^2 °C)	150
Filament cross-section	Circular
Filament diameter, w (mm)	0.4

a) Churchill correlation, natural convection.

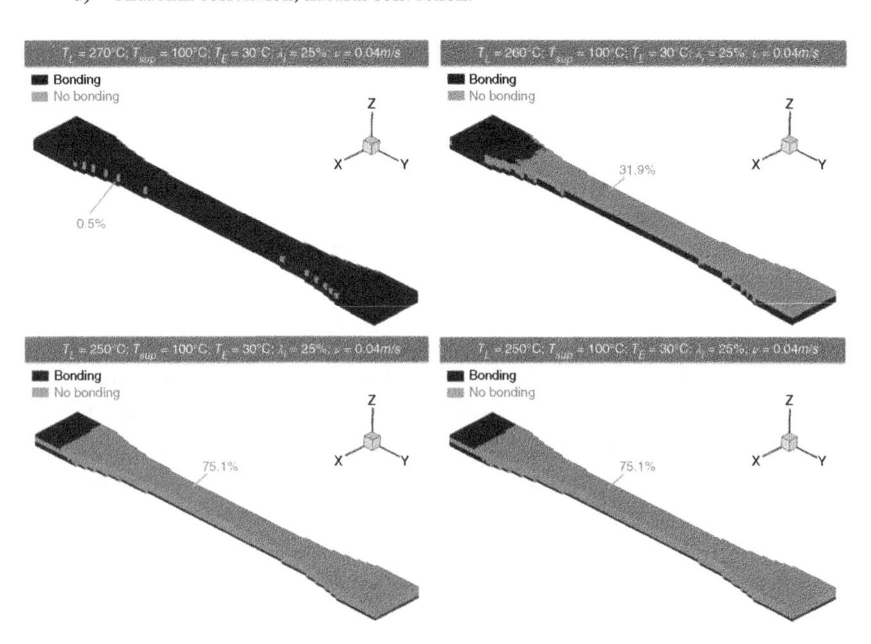

Figure 11.5 Effect of extrusion temperature on the extent of bonding.

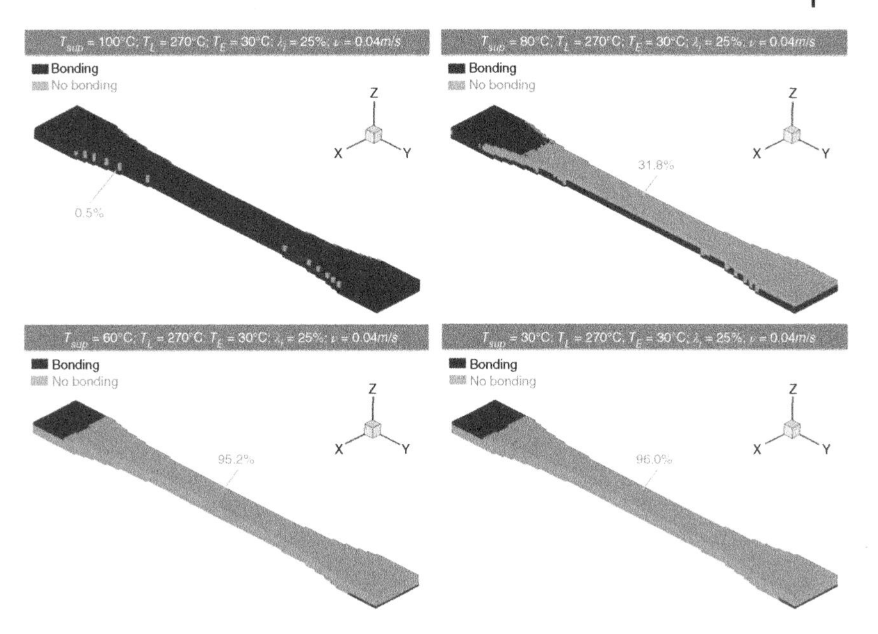

Figure 11.6 Effect of build platform temperature on the extent of bonding.

difficulties at the corner and curvature of the test specimen, which are often observed in practice. In fact, at these locations, the deposition strategy originates changes in the time periods for contact between adjacent threads, but simultaneously these are more exposed to the environment, which promotes faster cooling.

The temperature of the build platform is also an important process parameter since its value may induce faster cooling of the threads, particularly in the lower layers. Indeed, as shown in Figure 11.6, when the temperature of the build platform decreases from 100°C (which is approximately the glass transition temperature of ABS) to 30°C, the percentage of nonbonded or poorly bonded threads increases from 0.5 to 96%.

It is also interesting to estimate the role of the infill density on the final bonding achieved, as it influences the contact area between contiguous threads. Figure 11.7 shows the predictions of bonding for the part under study for different infill densities (70%; 80%; 90%; and 100%, corresponding to λ_i – the fraction of surface area in contact with another thread, of 17.5%, 20%, 22.5%, and 25%, respectively). The higher the infill density, the more frequent the contacts between threads occur; hence, the more efficient is the reheating above T_g of the older (and cooler) threads.

Using the combination of printing parameters selected, it was not possible to obtain a part with 100% bonding. This is mainly related to the deposition

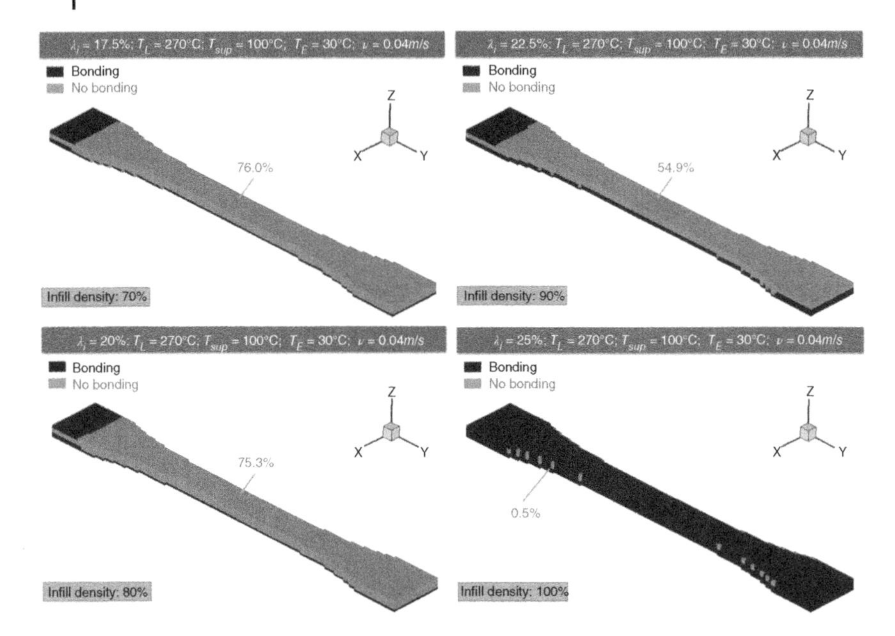

Figure 11.7 Effect of infill density λ_i on the extent of bonding. λ_i is the fraction of surface area in contact with another thread.

strategy adopted, which creates long times for contact of threads of different layers, due to the long length of the part. Although not exploited here, one solution would be to use a 3D printer with a build chamber and thus increase the environment temperature.

11.5 Conclusions

This chapter demonstrated the relevance of modeling the heat transfer phenomena developing during the deposition and cooling of the filament threads in the MEX process. The knowledge of the temperature evolution along the deposition allows a better understanding of the importance and role of the main process variables. This knowledge is also essential for predicting the bonding between adjacent filament threads, the mechanical properties and surface properties, or the warpage and dimensional tolerances of parts obtained by MEX.

Discretizing the volume to print into sufficiently small thread elements and solving the governing equations analytically or numerically using "birth" and "dead" elements can yield accurate predictions within affordable computational times. However, a precise characterization of the thermophysical properties of the polymer(s) and the correct choice of the values of the heat transfer coefficients are also needed.

References

1 Gosset, A., Barreiro-Villaverde, D., Becerra Permuy, J.C. et al. (2020). Abad López MJ Experimental and numerical investigation of the extrusion and deposition process of a poly(lactic acid) strand with fused deposition modeling. *Polymers* 12: 2885. https://doi.org/10.3390/polym12122885.

2 Xia, H., Lu, J., Dabiri, S., and Tryggvason, G. (2017). Fully resolved numerical simulations of fused deposition modeling. Part I—Fluid flow. *Rapid Prototyping Journal* 24: 463–476.

3 Xia, H., Lu, J., Dabiri, S., and Tryggvason, G. (2017). Fully resolved numerical simulations of fused deposition modeling. Part II—Solidification, Residual stresses and modeling of the nozzle. *Rapid Prototyping Journal* 24: 973–987.

4 Cattenone, A., Morganti, S., Alaimo, G., and Auricchio, F. (2019). Finite element analysis of additive manufacturing based on fused deposition modeling: distortions prediction and comparison with experimental data. *ASME Journal of Manufacturing Science and Engineering* 141 (1): 011010. https://doi.org/10.1115/1.4041626.

5 Yardimci, M.A. and Güçeri, S. (1996). Conceptual framework for the thermal process modelling of fused deposition. *Rapid Prototyping Journal* 2: 26–31. https://doi.org/10.1108/13552549610128206.

6 Thomas, J.P. and Rodriguez, J.F. (2000). Modeling the fracture strength between fused-deposition extruded roads. *Solid Freeform Fabrication Proceedings* 17–23. https://doi.org/10.26153/tsw/2054.

7 Rodriguez, J.F., Thomas, J.P., and Renaud, J.E. (2000). Characterization of the mesostructure of fused-deposition acrylonitrile-butadiene-styrene materials. *Rapid Prototyping Journal* 6: 175–185. https://doi.org/10.1108/13552540010337056.

8 Costa, S.F., Duarte, F.M., and Covas, J.A. (2017). Estimation of filament temperature and adhesion development in fused deposition techniques. *Journal of Materials Processing Technology* 245: 167–179. https://doi.org/10.1016/j.jmatprotec.2017.02.026.

9 Zhou, X., Hsieh, S.-J., and Sun, Y. (2017). Experimental and numerical investigation of the thermal behaviour of polylactic acid during the fused deposition process. *Virtual and Physical Prototyping* 12 (3): 221–233. https://doi.org/10.1080/17452759.2017.1317214.

10 Yin, J., Lu, C., Fu, J. et al. (2018). Interfacial bonding during multi-material fused deposition modeling (FDM) process due to inter-molecular diffusion. *Materials & Design* 150: 104–112. https://doi.org/10.1016/j.matdes.2018.04.029.

11 Zhang, Y. and Shapiro, V. (2018). Linear-time thermal simulation of as-manufactured fused deposition modeling components. *Journal of Manufacturing Science and Engineering* 140: 071002. https://doi.org/10.1115/1.4039556.

12 D'Amico, A. and Peterson, A.M. (2018). An adaptable FEA simulation of material extrusion additive manufacturing heat transfer in 3D. *Additive Manufacturing* 21: 422–430. https://doi.org/10.1016/j.addma.2018.0221.

13 Roy, M., Yavari, R., Zhou, C. et al. (2019). Prediction and experimental validation of part thermal history in the fused filament fabrication additive manufacturing process. *Journal of Manufacturing Science and Engineering* 141 (12): https://doi.org/10.1115/1.4045056.

14 Akhoundi, B. and Behravesh, A.H. (2019). Effect of filling pattern on the tensile and flexural mechanical properties of FDM 3D printed products. *Experimental Mechanics* 59: 883–897. https://doi.org/10.1007/s11340-018-00467-y.

15 Wang, P., Zou, B., and Ding, S. (2019). Modeling of surface roughness based on heat transfer considering diffusion among deposition filaments for FDM 3D printing heat-resistant resin. *Applied Thermal Engineering* 161: 114064. https://doi.org/10.1016/j.applthermaleng.2019.114064.

16 Lepoivre, A., Boyard, N., Levy, A., and Sobotka, V. (2020). Heat transfer and adhesion study for the FFF additive manufacturing process. *Procedia Manufacturing* 47: 948–955. https://doi.org/10.1016/j.promfg.2020.04.291.

17 Coogan, T.J. and Kazmer, D.O. (2020). Prediction of interlayer strength in material extrusion additive manufacturing. *Additive Manufacturing* 35: 101368. https://doi.org/10.1016/j.addma.2020.101368.

18 Gilmer, E.L., Anderegg, D., Gardner, J.M. et al. (2021). Temperature, diffusion, and stress modeling in filament extrusion additive manufacturing of polyetherimide: an examination of the influence of processing parameters and importance of modeling assumptions. *Additive Manufacturing* 48: Part A. https://doi.org/10.1016/j.addma.2021.102412.

19 Sinha, D., Lynch, S.P., and Meisel, N.A. (2021). Heat transfer simulation of material extrusion additive manufacturing to predict weld strength between layers. *Additive Manufacturing* 46: 102117. https://doi.org/10.1016/j.addma.2021.102117.

20 Xu, D., Zhang, Y., and Pigeonneau, F. (2021). Thermal analysis of the fused filament fabrication printing process: experimental and numerical investigations. *International Journal of Material Forming* 14: 763–776. https://doi.org/10.1007/s12289-020-01591-8.

21 Liparoti, S., Daniele, S., Romano, A. et al. (2021). Fused filament deposition of PLA: the role of interlayer adhesion in the mechanical performances. *Polymers* 13: 399. https://doi.org/10.3390/polym13030399.

22 Ramos, N., Mittermeier, C., and Kiendl, J. (2022). Experimental and numerical investigations on heat transfer in fused filament fabrication 3D-printed specimens. *The International Journal of Advanced Manufacturing Technology* 118: 1367–1381. https://doi.org/10.1007/s00170-021-07760-6.

23 Trofimov, A., Pavic, J., Pautard, S. et al. (2022). Experimentally validated modeling of the temperature distribution and the distortion during the fused filament fabrication process. *Additive Manufacturing* 54: 102693. https://doi.org/10.1016/j.addma.2022.102693.

24 Costa, S.F., Duarte, F.M., and Covas, J.A. (2014). Thermal conditions affecting heat transfer in FDM/FFE: a contribution towards the numerical modelling of the process. *Virtual and Physical Prototyping* 10: 1–12. https://doi.org/10.1080/17452759.2014.984042.

25 Duarte, F.M., Covas, J.A., and Costa, S.F. (2022). Predicting the effect of build orientation and process temperatures on the performance of parts made by fused filament fabrication. *Rapid Prototyping Journal* 28: 704–715. https://doi.org/10.1108/RPJ-04-2021-0084.

26 Yang, F. and Pitchumani, R. (2002). Healing of thermoplastic polymers at an interface under nonisothermal conditions. *Macromolecules* 35: 3213–3224. https://doi.org/10.1021/ma010858o.

27 Costa, S.F., Duarte, F.M., and Covas, J.A. (2021). The effect of a phase change on the temperature evolution during the deposition stage in fused filament fabrication. *Computers* 10: 19. https://doi.org/10.3390/computers10020019.

28 Vanaei, H.R. et al. (2021). Experimental study of PLA thermal behavior during fused filament fabrication. *Journal of Applied Polymer Science* 138: 49747. https://doi.org/10.1002/app.49747.

12

In-Process Temperature Monitoring in 3D Printing

Saeedeh Vanaei[1] and Michael Deligant[2]

[1] *Department of Mechanical, Industrial and Manufacturing Engineering, University of Toledo, Toledo, OH, 43606, USA*
[2] *Arts et Métiers Institute of Technology, CNAM, LIFSE, HESAM University, 75013, Paris, France*

12.1 Introduction

Additive manufacturing (AM), also known as 3D printing, is a process of joining layers to create a 3D object from 3D models. There have been lots of works in the last two decades due to its ability to reduce cost and time as well as enable the construction of complex 3D geometries [1]. There are various AM techniques based on different mechanisms, including layer-by-layer printing, stereolithography (SLA), selective laser sintering (SLS), laminated object manufacturing (LOM), fused filament fabrication (FFF), and others [2].

In terms of different AM techniques, FFF has garnered significant attention owing to its capability to produce functional prototypes using various polymers, thermoplastics, and composites. FFF, also known as fused deposition modeling (FDM) [3], was used for prototyping purposes, allowing the fabrication of complex geometries, and easily operable parts. However, despite the aforementioned advantages of FFF, the quality of the constructed objects remains a critical aspect to address [4].

One of the most significant issues in 3D printing is that the bonding location becomes the point of failure. Due to the layer-by-layer deposition mechanism, the process involves depositing a hot layer onto previously deposited layers that are solidifying. This results in temperature differences between the deposited layers, with cooling and reheating occurring. As the adhesion of these layers relies on thermal properties, the temperature profile becomes a crucial parameter to consider in the analysis of layer adhesion [5].

Industrial Strategies and Solutions for 3D Printing: Applications and Optimization,
First Edition. Edited by Hamid Reza Vanaei, Sofiane Khelladi, and Abbas Tcharkhtchi.

Over the years, researchers and practitioners have recognized that heat transfer plays a critical role in the 3D-printing process and directly affects the final quality and performance of printed objects. Heat transfer refers to the movement of thermal energy between different components or regions within a system. In the context of 3D printing, it involves the transfer of heat between the printing material and the surrounding environment as well as within the material itself during the printing process [6]. The chapter seeks to persuade readers of the significance of understanding and optimizing heat transfer in 3D printing. By doing so, it aims to underscore the impact that heat transfer can have on the strength and overall characteristics of 3D-printed parts. The strength of a printed object is crucial, as it determines its ability to withstand external forces, stresses, or mechanical loads without failure or deformation.

12.2 Heat Transfer in 3D Printing

The layer-by-layer mechanism in FFF is the result of extruding filaments through a liquefier, which itself moves in three dimensions through successive layers in the x–y planes along the z direction. During 3D printing, a filament, often made of a thermoplastic material, is heated until it reaches a molten state. This molten filament is then extruded through a nozzle and deposited layer by layer to build the desired object. As the freshly deposited filament begins to cool down, it solidifies and undergoes a phase transition from a molten state to a solid state. This temperature increase can be significant enough to reach or surpass the glass transition temperature (T_g) for amorphous materials or the crystallization temperature (T_c) for semicrystalline materials. The T_g represents the temperature at which amorphous materials transition from a hard and brittle state to a more pliable and rubbery state. On the other hand, the T_c represents the temperature at which semicrystalline materials undergo a transition from a disordered amorphous state to a more organized crystalline structure. Reaching or exceeding the T_g or T_c at the interfaces of the contacting filaments is a critical point in the 3D-printing process. It signifies that the material in those regions has experienced a significant change in its internal structure due to the reheating. This can affect the overall properties and characteristics of the printed object, including its mechanical strength, dimensional stability, and thermal behavior. This temperature threshold is necessary for proper adhesion at the filament interfaces. Consequently, each filament should be sufficiently heated during deposition to prevent deformation caused by the weight and gravity of the filaments in subsequent layers. There are several heat transfer mechanisms involved during FFF, outlined as follows [7]:

(1) The heat generated by the liquefier during the 3D-printing process.
(2) The cooling of the filaments through convection with the surrounding air, where the heat transfer coefficient (h_{conv}) plays a significant role.
(3) The transfer of heat between neighboring filaments, which is influenced by their conductance and the order in which they are deposited.
(4) The heat transmitted through the support plate via conduction, which is controlled by the thermal contact conductance and contact area.
(5) Loss of heat through radiation, including radiation between the filament and its surroundings as well as radiation between adjacent filaments.
(6) The heat generated from the exothermal crystallization process in semicrystalline polymers.

There have been several studies on predicting the thermal history of filaments during deposition. Costa et al. [8] proposed an analytical solution for transient heat transfer during filament deposition, considering filament contacts. Although they neglected axial and radial heat conduction, they have recently investigated the contribution of various thermal phenomena present in the FFF process [9].

Bellehumeur et al. [10] assumed a uniform temperature profile and introduced the lumped capacity model for the filament interfaces. They simplified their model to a 1D transient heat transfer model to predict the cooling and cyclic temperature profile of the deposited layers. However, despite the simplicity of their approach, the simulation does not account for complex geometries. Their approach suffers from a significant limitation in terms of experimental validation and a notable disparity between the results obtained and real-world conditions. In contrast, Yardimici et al. [11] put forward a 1D thermal model in their research that takes into account thermal interaction with the surroundings and between the filaments that are deposited.

In the last decade, with the expiration of the Stratasys FDM™ patent [12], more research opportunities have emerged in the field of optimization. The primary reason for this is the widespread availability of open-source 3D printers, enabling researchers to explore different optimization possibilities. In a study by Peng et al. [13], temperature sensors were utilized to analyze temperature fluctuations throughout the extrusion process. The research revealed that an increase in print speed leads to a departure from isothermal flow. Their findings indicate that 3D printing is a non-isothermal process and highlight the crucial role of temperature variation, which should be controlled for optimization purposes.

In addition to other works, there have been similar experimental studies that employed a specific experimental setup to examine the relationship between feed rate and feeding force at different liquefier temperatures [14]. These studies revealed that the liquefier temperature serves as a limiting factor in the process. Similarly, Vaes et al. [15] utilized an infrared (IR) sensor with a thermal camera to analyze temperature variations parallel to filament deposition and enable in-process monitoring of the temperature profile. Cattenone et al. [16] employed the finite element method (FEM) to predict distortion in 3D-printed fabricated parts. They further validated their simulations experimentally and concluded that the mechanical properties of the 3D-printed parts are significantly influenced by temperature variations in the deposited layers during the deposition process.

A particularly fascinating research endeavor in the realm of thermal analysis for 3D-printed components is the study carried out by Seppala and Migler [17]. They employed an IR camera to examine the temperature distribution in the vicinity of the printing region. Similarly, D'Amico and Peterson [18] utilized the FEM to simulate heat transfer throughout the FFF process. When considering the measurement mechanism of IR cameras, which is based on in-process monitoring of the external surface of printed parts, it is worth noting that while their model was validated, the experimental technique employed provides a global temperature profile. As a result, the developed model relies on these obtained results.

In more recent studies, Costa et al. introduced a heat transfer model that integrates the transient heat conduction occurring between the filaments and/or the platform on which they are built [8, 19]. Their research revealed that the temperature distribution of the filaments is notably impacted by the physical contact between them. The key point of their study lies in the direct implementation of physical contacts between the layers, showcasing the importance of local in-process monitoring in this process and its impact on temperature variations within the layers.

Although they used an IR camera to validate their model, the key point of their study was the inclusion of physical contact. The comparison of results highlighted the significant impact of the filaments' interfaces due to this consideration.

The chapter, in this book, offers an extensive examination of how the heat transfer process in 3D printing is affected by essential parameters such as the temperature of the liquefier, platform, and print speed. It provides a thorough discussion of the influence of these parameters on the heat transfer dynamics within the 3D-printing process. In addition to the introduction and fundamentals of heat transfer, the book covers the advantages and limitations of local and global in-process monitoring techniques as well as various numerical models proposed by researchers.

12.3 The Impact of Cyclic Temperature Profile in 3D-Printing Process

The role of heat transfer, particularly the heat sources, is essential in the 3D-printing process, and the quality of the final parts is significantly influenced by temperature variation. It is important to note that 3D printing is a non-isothermal process, and heat sources play a crucial role in this phenomenon. Process variables such as liquefier temperature, platform temperature, and print speed directly affect the viscosity of the filaments during deposition and solidification. Considering the complexity of the FFF process and its non-isothermal characteristics as well as the impact of process variables, it becomes evident that the melt rheology and strength of the final parts are affected. Viscosity, being a temperature-dependent parameter, necessitates an investigation into heat transfer and temperature variation of deposited layers. Due to this fact, it is crucial to evaluate the temperature variation at the interface, as adhesion is a thermally driven process phenomenon. From the very beginning, researchers have dedicated their efforts to studying heat transfer in FFF 3D-printed parts. They have presented numerous experimental and analytical works to better understand the significance of temperature variation and to gain a clearer perspective on the state of the art. Open-source software and hardware refer to technologies that are developed collaboratively and made freely available to the public, allowing anyone to study, modify, and distribute them. These open-source solutions have gained popularity in various industries due to their accessibility, flexibility, and ability to foster innovation. The expiration of the Stratasys FDM™ patent, combined with the availability of open-source software and hardware, has led to a surge in progressive development in the field of 3D printing. With the patent no longer limiting the use of FDM technology, more companies and individuals have been able to explore and experiment with it. Open-source initiatives have facilitated collaboration and knowledge sharing among developers, enabling them to build upon existing designs and create new and innovative applications. As a result, the availability of open-source software and hardware, coupled with the expiration of the Stratasys FDM™ patent, has contributed to the rapid advancement of 3D-printing technology. This progress has encompassed various aspects, such as improved printer designs, enhanced print quality, expanded material compatibility, and the development of novel applications in fields like healthcare, engineering, manufacturing, and more. Additionally, there has been significant attention given to the measurement of filament temperature using various experimental methods. Common techniques include the implementation of IR cameras or thermocouples for in-process monitoring of the temperature profile parallel to the layer deposition. It is important to note that each technique has its own advantages and limitations. To be more precise, the application of an IR camera provides the temperature profile of the external surface, which represents a global approach. On the other hand, the implementation of thermocouples in between the deposited layers allows for recording the temperature profile at the interfaces of the deposited layers, which represents a local approach. Consequently, it is important to distinguish the advantages and limitations of each approach, and these features will be discussed in detail in this section.

A thermal model refers to a mathematical or computational representation of the thermal behavior or characteristics of a system or object. It is commonly used to analyze and predict temperature distribution, heat transfer, and other thermal phenomena. Developing a thermal model typically involves a series of steps and considerations to accurately represent real-world thermal processes. By presenting the thermal model's description and summarizing it in Table 12.1, the reader can quickly refer to the table to gain a high-level understanding of the model's key features. The table provides a concise overview of the model's formulation, allowing readers to grasp the main elements without needing to go through the entire detailed description in the chapter. Yardimci and Guceri [21] predicted the temperature profile while examining a ceramic-filled thermoplastic material. They successfully predicted the layer adhesion, although the limited availability of experimental data posed a challenge. In a subsequent study, Yardimci et al. [11] employed the FEM to simulate the temperature profile of filaments. They discovered that the absence of deposited layers resulted in a gradual decrease in the cooling rate. Thomas and Rodriguez [22] proposed a thermal model, which likely involves mathematical or computational calculations, to analyze and quantify the interlayer fracture strength. This thermal model considers the thermal behavior and associated stresses during the printing process, taking into account factors such as heat transfer, thermal gradients, and cooling rates. By incorporating these thermal aspects into the model, the authors aim to establish a correlation between the thermal conditions and the resulting interlayer fracture strength. Zhang and Chou's FEM model [23] considered the crucial aspect of full contact between deposited layers. In AM, achieving proper bonding between layers is essential for the overall strength and integrity of the

Table 12.1 Summary of the most common modeling related to heat transfer in FFF.

Modeling methodology	Description	Obtained results
Finite element method	– Applicable for both metallic or polymeric parts	– Sequence of layers is an important issue – Heat transfer affected by advection
Finite Element Method using ANSYS	– Including both convection and conduction – Residual stress and dimensional accuracy affected by temperature	– Cyclic cooling reheating temperature profile – Residual stress affected by this temperature profile – Dimensional accuracy affected by temperature
Analytical heat transfer modeling	– Single road width geometry – Rectangular road cross-section area – Neglection of heat transfer along road length	– Road cross-section area is an indicator in cooling process – The higher the h_{conv}, the higher the difference between center and edge temperature
Finite element method using ABAQUS	– Including convection, conduction, and radiation – Optimization using different values for thermal conductivity	– Possibility of neglecting the axial conduction – Neglection of radiation from a specific amount of h_{conv}
Finite volume method	– Spatial and temporal discretization – Conduction and natural convection – Continuously moving boundary	– The higher the h_{conv}, the higher the bonding – Less bonding near edges

Source: [20]/MDPI/CCBY 4.0/Public domain.

printed object. By considering full contact, the model accounts for the interaction and influence of neighboring layers on each other during the printing process. By employing FEM simulations, Costa et al. [9] could quantitatively assess how these heat transfer mechanisms affect temperature distribution, heat flux, and overall thermal behavior in their specific context of study. The simulations likely involved mathematical models, boundary conditions, and material properties to accurately represent the physical phenomena. However, in all the aforementioned works, the lack of experimental data limited the scope of experimental validation. On the other hand, the method chosen for experimental validation plays a crucial role, highlighting the importance of a well-defined experimental setup. Through the proposal of different modeling and simulation approaches, researchers have obtained various characteristics related to heat transfer.

In this section, various concepts related to the heat transfer of 3D-printed parts are presented, categorized as follows: the in-process monitoring of temperature profiles is explained, along with the different techniques that are available. Additionally, the advantages and limitations of each technique are discussed to facilitate a clear distinction for further objectives. Special attention has been given to discussing the most recent studies in order to apply their findings for optimization purposes.

12.3.1 In-Process Monitoring of Temperature Variation in 3D-Printing Process

Given the significance of heat transfer in the 3D-printing process, temperature measurement, and particularly in-process monitoring, is essential. The primary objective of this chapter is to focus on this aspect. Numerous experimental, numerical, and combined experimental–numerical studies have been considered to thoroughly examine the temperature profile of filaments during the deposition process.

The 3D-printing process involves the layer-by-layer deposition of material to create a 3D object. Understanding the temperature dynamics during this process is crucial for optimizing print quality, material behavior, and overall performance. To achieve a comprehensive understanding of temperature variations, the process has been divided into global monitoring and local monitoring. Global monitoring refers to temperature recording on the external surface of deposited layers. This approach involves measuring and analyzing temperature changes on the outer surface of the printed object. Global monitoring provides an overall perspective on the temperature distribution and behavior during the printing process. It helps in identifying macroscopic trends and patterns, and it is particularly useful for evaluating the impact of different printing parameters or environmental conditions on the temperature of the printed object as a whole. On the other hand, local monitoring involves temperature recording at the interface of adjacent layers. This approach focuses on measuring and analyzing temperature variations at the contact points between layers. By monitoring the temperature at the interface, researchers gain insights into the bonding behavior between layers and the potential presence of temperature-related defects, such as insufficient fusion or delamination. Local monitoring provides a more detailed understanding of the thermal characteristics and interlayer interactions during the printing process.

The authors of the mentioned work have dedicated their attention to various studies conducted in both global and local monitoring categories. By examining the existing research, they aim to differentiate and compare the advantages and limitations of the works conducted in these two areas. This comparison helps in identifying the strengths and weaknesses of different monitoring techniques, experimental setups, and data analysis methods. It allows researchers and practitioners to make informed decisions when selecting an appropriate temperature monitoring approach based on their specific goals and requirements. Overall, by dividing the 3D-printing process into global monitoring and local monitoring and by reviewing studies in both categories, the authors aim to enhance the understanding of temperature dynamics during the printing process. This analysis contributes to the knowledge base in the field, provides insights into the advantages and limitations of different monitoring approaches, and ultimately assists in improving the quality and reliability of 3D-printed objects.

12.3.1.1 Global Monitoring – Temperature Recording on the External Surface of Deposited Layers

One of the most common techniques for temperature monitoring of the external surface of 3D-printed parts is the implementation of IR cameras [24]. Researchers take advantage of this approach to gather data.

In a study by Seppala and Migler [17], an IR camera was employed to measure the spatial and temporal temperature profile of filaments under various printing conditions, with a focus on the weld zone and conducting in-process monitoring of filament temperature profile. They estimated the sublayer heating during filament deposition. In another work, they expanded on their previous research by investigating weld formation during filament deposition, using an experimental framework that incorporated thermal and rheological characteristics [25]. They conducted a parametric study by considering print speed and liquefier as heat sources, measuring the temperature profile of filaments during deposition. While they noted that an increase in weld time led to an increase in weld strength, further studies are required for IR thermography, as the recording mechanism differs from weld strength measurements.

Rudolph et al. [26] presented a numerical approach utilizing Python™ and ANSYS® for tool path generation and heat transfer simulations. To validate their proposed approach, they employed an IR camera to assess the cooling and reheating process during filament deposition. While they argued that radiation is a crucial factor in filament heat transfer, they did not consider the recording mechanism of the measurements based on their findings.

In another study, El Moment et al. [27] employed a 3D-thermomechanical model to assess the temperature profile and residual stress field. The notable feature of their developed model is the inclusion of various material characteristics, including thermal conductivity, yield stress, and Young's modulus, which are temperature-dependent phenomena. The findings revealed that temperature variation leads to higher stress concentration and, consequently, increased residual thermal stress, thereby elevating the probability of failure in 3D-printed parts.

Compton et al. [28] proposed an experimental–numerical approach to evaluate temperature variation in large-scale polymer composite materials. They utilized an IR camera to record the global temperature profile of filaments and validated the obtained results using a 1D finite difference method for heat transfer modeling of the corresponding temperature profile. The authors observed that the platform temperature directly affects the cooling rate of the deposited layers. Furthermore, they noticed discrepancies between the onset of peaks in the temperature profiles obtained from the IR camera and their developed model, indicating limitations of the IR camera, such as the overestimation of recorded data. In addition to the aforementioned research studies, combined experimental–numerical in-process monitoring of temperature profiles has been widely considered and continues to evolve, as it contributes to conducting more comprehensive studies. It is worth noting that the authors specifically reported an overestimation of the results in the recorded temperature profile compared to the numerical approaches they employed for experimental validation, as mentioned earlier.

The effect of process parameters on the temperature variation at the interface of adjacent layers was investigated by Kuznetsov et al. [29] using tube-shaped 3D-printed parts with a rectangular cross-section. Despite their objective of studying the characteristics at the interface of deposited layers, they employed an IR camera and determined the average temperature distribution on the external surface of the deposited layers. It is important to note that the temperature evaluation mechanism they used was contrary to the objective of their study, and, therefore, their conclusions could not be used as a criterion for evaluating the adhesion of the deposited layers.

Basgul et al. [30] proposed a non-isothermal healing model to measure the temperature profile of a vertical wall. However, they utilized an IR camera to experimentally record the temperature profile of the deposited layers. Unfortunately, they observed poor agreement between the obtained experimental and numerical results.

In all of the abovementioned works, the employed experimental approach for temperature measurement was the utilization of IR cameras, specifically for measuring the temperature profile at the external surface of the deposited layers. To provide a better understanding, a summary of the in-process monitoring of temperature profiles using IR cameras has been presented in Table 12.2, highlighting the most important outcomes of each research work.

12.3.1.2 Local Monitoring – Temperature Recording at the Interfaces of Adjacent Layers

In-process monitoring of the temperature profile of deposited layers needs to be precise and efficient in order to capture the temperature variation at the interfaces accurately. One crucial aspect of in-process monitoring is the ability to quickly record the cooling and reheating peaks that occur at the contacts between the deposited layers. To address this, Kousiatza and Karalekas [31] conducted a real-time study, where they simultaneously measured the strain and temperature profiles during FFF process. They integrated fiber Bragg grating (FBG) sensors to record the solidification-induced strain levels and assess the impact of temperature profiles on these characteristics. They proposed an experimental–numerical approach using thermocouples to record the filament's temperature at the interface of adjacent layers, as depicted in Figure 12.1 [32]. The results of both studies demonstrated a strong agreement between the experimental temperature profiles and the predicted results. Furthermore, the peak values observed during the reheating of previously deposited layers emphasized the value of local measurements in understanding the process.

Similarly, Xu et al. [33] utilized T-type thermocouples during the deposition of filaments to measure the temperature variation of the deposited layers when constructing a vertical wall. By employing their previously developed model, the authors claimed a good agreement between the predicted results and the recorded experimental data. However, it should be noted that the process of pausing to position the thermocouples could introduce variations in the recorded experimental results, which poses a challenge in their analysis.

To summarize, this section has presented the in-process monitoring of the temperature profile of deposited layers during the FFF process. In order to provide a clearer understanding, both local and global in-process monitoring techniques have been summarized in Table 12.3, highlighting the materials used and the obtained results.

Table 12.2 Summary of in-process monitoring of temperature variation using IR camera.

Research	Obtained results
Infrared thermography of welding zone, in-process monitoring of temperature profile	– Sudden drop in temperature profile – Quick decrease in estimated weld temperature
Infrared thermography of filament's temperature history at their interfaces	– 13% difference in FEM and experimental results – Longest thermal diffusion at specific parameters
Thermal analysis of large-scale thermoplastic polymer composites	– Excellent agreement of FDM heat transfer model and experiment results – Temperature of the top layer: indicator for degree of warping and cracking
Weld formation during FFF	– Developing of a framework, including thermal history and mechanical properties
Filament temperature distribution in the stand-off gap between liquefier and platform	– Good agreement of modeling and experimental data – Developing a framework of heat transfer and process optimization
Effect of temperature field on mechanical strength	– Platform temperature affects the cooling and reheating – Radiation should be considered in temperature investigations – Presence of voids affects the cooling and reheating
Evaluation of thermal properties of 3D spacer	– Air gap can increase insulation efficiency
Temperature and residual stress modeling	– Temperature difference of simulation and experimental: <5% – Stress variation as a result of decrease in temperature
Influence of forced-air cooling	– The higher the airflow, the lower the mechanical strength
Thermography-based in-process monitoring of temperature profile	– Poor agreement of obtained results from model and experimental recorded data
Influence of temperature-related parameters on strength of 3D-printed parts	– Temperature plays an important role on inter-layer bonding

Table 12.2 (Continued)

Research	Obtained results
Heat transfer and adhesion study	– Good agreement of recorded and predicted temperature profile – Obtained results are the basis for coalescence phenomena
Heat transfer and interfacial bonding strength	– Upper deposited layers showed higher temperature – Liquefier temperature has significance effect on layer healing – Good agreement between results of model and experimental recorded data

Source: [20]/MDPI/CCBY 4.0/Public domain.

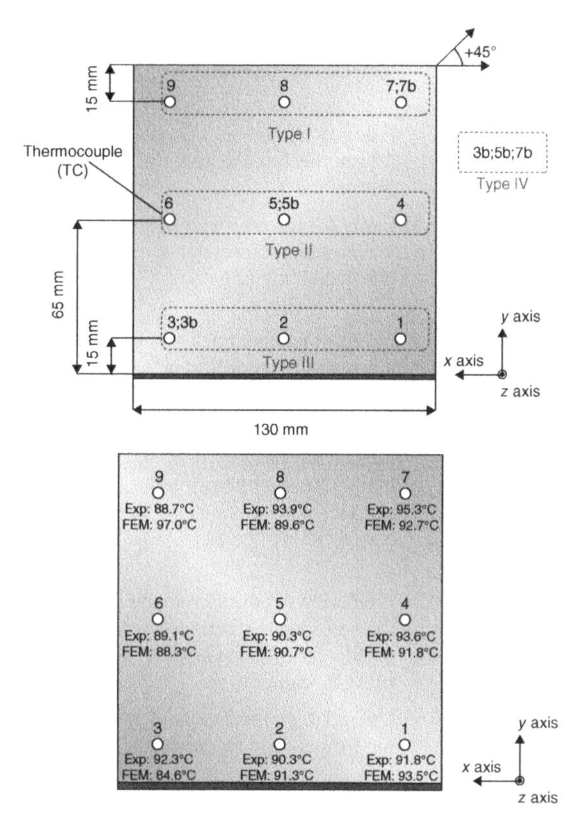

Figure 12.1 Schematic representation of location of the thermocouples and the experimental–numerical obtained temperature values (Source: [32]/MDPI/CCBY 4.0/Public domain).

Table 12.3 Summary of in-process monitoring of temperature variation techniques.

Material	Description	Results
ABS	Simultaneous monitoring of temperature and strain using fiber Bragg grating sensor	Process optimization by correlation of temperature and residual stress
CF-ABS	Measuring the thermal evolution of composite	Evaluation of thermal stress evolution, warping, and fracture initiation
PLA	Measuring the temperature profile of deposited layers	High difference between the model and recorded results
PLA	Using IR camera for temperature recording	Different cooling rates in different sections correspond to their mechanical properties
ABS	Using K-type thermocouples	Good agreement with the developed model
TPU	Using IR camera for temperature recording	Evaluation of thermal properties of 3D spacer technical material
PA12	Using IR camera as a part of thermomechanical analysis of printed parts	Modeling of temperature variation and residual stress (5%)
TPU-ABS	Using K-type thermocouples for implementation on interfacial bonding	Understanding of interfacial bonding mechanism to improve the mechanical properties
PEEK	Using IR camera for temperature recording in non-isothermal healing model for interfacial bonding	Liquefier temperature has high impact on layer healing, 100% healing at high platform temperature
PLA	Measuring surface temperature of the printed samples	Temperature recording permits the strength approximation of the interlayer bonding of the material
ABS-PEEK	Prediction of adhesion using the recorded temperature profile of deposited layers using IR camera	Implementing the obtained results in rheological characteristics
ABS	Using T-type thermocouples for temperature recordings	Developing a model, using the obtained results for prediction of bonding quality

Source: Adapted from Ref. [20].

12.4 Advantages and Disadvantages of Global–Local In-Process Monitoring

As explained earlier, both global and local in-process monitoring techniques for temperature profiles in the FFF process have their own advantages and limitations. While an IR camera may have limited scanning quality in complex geometries, thermocouples can only be fixed at a limited number of points on a geometry. In order to address the variety of approaches in real-time monitoring, Vanaei et al. [34] conducted a comprehensive experimental comparison of these techniques. By simultaneously implementing very small K-type thermocouples and an IR camera during the filament deposition without pausing the process, the authors were able to identify the recorded temperature profile obtained.

Given the aforementioned observations and the presented studies, the advantages and limitations of these approaches can be listed as follows:

- Global approach is capable of recording temperature in complex geometries, while the local approach provides the temperature profile at the interface of adjacent filaments.
- In the global approach, there is an overestimation of peak values, while in the local approach, the peak values correspond to the interfaces' temperature.
- The local approach and the obtained temperature profile are useful for predicting adhesion and interlayer bonding, while the global approach gives the temperature profile on the external surface of the deposited layers.
- The local approach and its obtained results are useful for rheological characteristics as viscosity is a temperature-dependent parameter, and the material flow at the interface plays an important role in this regard.

12.5 Summary and Outlook

The advantages of 3D-printing/FFF process in producing complex geometries provide the possibility of fabricating different polymer/polymer-based composite materials. Due to the presence of multiple heat sources and various heat transfer mechanisms, FFF is considered a thermally driven process. Therefore, heat transfer plays a crucial role in determining the temperature history of the deposited layers. In-process monitoring of the temperature profile is essential for gaining a better understanding of the temperature evolution during the deposition process. This enables further investigations such as optimization purposes and the prediction of interlayer bonding and adhesion of the layers.

In this chapter, we have provided a comprehensive summary of the most important works on global and local in-process monitoring of temperature profiles, aiming to distinguish the differences between the two approaches. The utilization of IR cameras as a global approach and various types of thermocouples as a local approach have been extensively discussed, highlighting their significance as measurement devices in research studies. Special attention has been given to emphasizing the techniques of global and local in-process monitoring and elucidating the advantages and limitations of each approach. Through the use of experimental, numerical, and combined experimental–numerical approaches, substantial progress has been achieved in the field of in-process monitoring of temperature profiles. It has been demonstrated that obtaining a precise temperature profile at the interface of the deposited layers is crucial for optimizing the process and predicting interlayer bonding. Closing the gap between recorded data and the actual temperature variations would significantly contribute to achieving higher quality in the fabrication of 3D-printed parts.

References

1 Hiemenz, J. (2011). *3D Printing with FDM: How it Works*, vol. 1, 1–5. Stratasys Inc.

2 Kruth, J.-P. (1991). Material incress manufacturing by rapid prototyping techniques. *CIRP Annals* 40 (2): 603–614.

3 Crump, S.S. (1991). Fast, precise, safe prototypes with FDM. *ASME, PED* 50: 53–60.

4 Magri, A.E., El Mabrouk, K., Vaudreuil, S. et al. (2021). Mechanical properties of CF-reinforced PLA parts manufactured by fused deposition modeling. *Journal of Thermoplastic Composite Materials* 34 (5): 581–595.

5 Vanaei, H., Raissi, K., Deligant, M. et al. (2020). Toward the understanding of temperature effect on bonding strength, dimensions and geometry of 3D-printed parts. *Journal of Materials Science* 55 (29): 14677–14689.

6 Dizon, J.R.C., Espera, A.H. Jr., Chen, Q. et al. (2018). Mechanical characterization of 3D-printed polymers. *Additive Manufacturing* 20: 44–67.

7 Adam, L., Lietaer, O., Mathieu, S. et al. (2020). Numerical simulation of additive manufacturing of polymers and polymer-based composites. In: *Structure and Properties of Additive Manufactured Polymer Components*, 115–146. Elsevier.

8 Costa, S., Duarte, F., and Covas, J. (2008). Towards modelling of Free Form Extrusion: analytical solution of transient heat transfer. *International Journal of Material Forming* 1 (1): 703–706.

9 Costa, S., Duarte, F., and Covas, J. (2015). Thermal conditions affecting heat transfer in FDM/FFE: a contribution towards the numerical modelling of the process: this paper investigates convection, conduction and radiation phenomena in the filament deposition process. *Virtual and Physical Prototyping* 10 (1): 35–46.

10 Bellehumeur, C., Li, L., Sun, Q. et al. (2004). Modeling of bond formation between polymer filaments in the fused deposition modeling process. *Journal of Manufacturing Processes* 6 (2): 170–178.

11 AtifYardimci, M., Hattori, T., Guceri, S.I. et al. (1997). Thermal analysis of fused deposition. In: *1997 International Solid Freeform Fabrication Symposium*.

12 Crump, S.S. (1994). *Modeling Apparatus for Three-Dimensional Objects*. Google Patents.

13 Peng, F., Vogt, B.D., and Cakmak, M. (2018). Complex flow and temperature history during melt extrusion in material extrusion additive manufacturing. *Additive Manufacturing* 22: 197–206.

14 Serdeczny, M.P., Comminal, R., Pedersen, D.B. et al. (2020). Experimental and analytical study of the polymer melt flow through the hot-end in material extrusion additive manufacturing. *Additive Manufacturing* 32: 100997.

15 Vaes, D., Coppens, M., Goderis, B. et al. (2019). Assessment of crystallinity development during fused filament fabrication through fast scanning chip calorimetry. *Applied Sciences* 9 (13): 2676.

16 Cattenone, A., Morganti, S., and Alaimo, G. (2019). Finite element analysis of additive manufacturing based on fused deposition modeling: distortions prediction and comparison with experimental data. *Journal of Manufacturing Science and Engineering* 141 (1).

17 Seppala, J.E. and Migler, K.D. (2016). Infrared thermography of welding zones produced by polymer extrusion additive manufacturing. *Additive Manufacturing* 12: 71–76.

18 D'Amico, A. and Peterson, A.M. (2018). An adaptable FEA simulation of material extrusion additive manufacturing heat transfer in 3D. *Additive Manufacturing* 21: 422–430.

19 Costa, S., Duarte, F., and Covas, J. (2017). Estimation of filament temperature and adhesion development in fused deposition techniques. *Journal of Materials Processing Technology* 245: 167–179.

20 Vanaei, H.R., Shirinbayan, M., and Deligant, M. (2021). In-process monitoring of temperature evolution during fused filament fabrication: a journey from numerical to experimental approaches. *Thermo* 1 (3): 332–360.

21 Yardimci, M.A. and Güçeri, S. (1996). Conceptual framework for the thermal process modelling of fused deposition. *Rapid Prototyping Journal.*

22 Thomas, J. and Rodríguez, J. (2000). Modeling the fracture strength between fused-deposition extruded roads 16. In: *2000 International Solid Freeform Fabrication Symposium.*

23 Zhang, Y. and Chou, Y. (2006). Three-dimensional finite element analysis simulations of the fused deposition modelling process. *Proceedings of the Institution of Mechanical Engineers, Part B: Journal of Engineering Manufacture* 220 (10): 1663–1671.

24 Zhou, X. and Hsieh, S.-J. (2017). Thermal analysis of fused deposition modeling process using infrared thermography imaging and finite element modeling. In: *Thermosense: Thermal Infrared Applications XXXIX.* International Society for Optics and Photonics.

25 Seppala, J.E., Han, S.H., Hillgartner, K.E. et al. (2017). Weld formation during material extrusion additive manufacturing. *Soft Matter* 13 (38): 6761–6769.

26 Rudolph, N., Chen, J., and Dick, T. (2019). Understanding the temperature field in fused filament fabrication for enhanced mechanical part performance. In: *AIP Conference Proceedings.* AIP Publishing LLC.

27 El Moumen, A., Tarfaoui, M., and Lafdi, K. (2019). Modelling of the temperature and residual stress fields during 3D printing of polymer composites. *The International Journal of Advanced Manufacturing Technology* 104 (5): 1661–1676.

28 Compton, B.G., Post, B.K., Duty, C.E. et al. (2017). Thermal analysis of additive manufacturing of large-scale thermoplastic polymer composites. *Additive Manufacturing* 17: 77–86.

29 Kuznetsov, V.E., Solonin, A.N., Tavitov, A. et al. (2020). Increasing strength of FFF three-dimensional printed parts by influencing on temperature-related parameters of the process. *Rapid Prototyping Journal.*

30 Basgul, C., Thieringer, F.M., and Kurtz, S.M. (2021). Heat transfer-based non-isothermal healing model for the interfacial bonding strength of fused filament fabricated polyetheretherketone. *Additive Manufacturing* 102097.

31 Kousiatza, C. and Karalekas, D. (2016). In-situ monitoring of strain and temperature distributions during fused deposition modeling process. *Materials & Design* 97: 400–406.

32 Kousiatza, C., Chatzidai, N., and Karalekas, D. (2017). Temperature mapping of 3D printed polymer plates: experimental and numerical study. *Sensors* 17 (3): 456.

33 Xu, D., Zhang, Y., and Pigeonneau, F. (2020). Thermal analysis of the fused filament fabrication printing process: experimental and numerical investigations. *International Journal of Material Forming* 1–14.

34 Vanaei, H., Deligant, M., Shirinbayan, M. et al. (2021). A comparative in-process monitoring of temperature profile in fused filament fabrication. *Polymer Engineering & Science* 61 (1): 68–76.

13

Optimizing the Controlling Factors and Characteristics of 3D-printed Parts

Anouar El Magri and Sébastien Vaudreuil

Euromed Polytechnic School, Euromed Research Center, Euromed University of Fes, Fès, Morocco

13.1 Introduction

Over the past years, additive manufacturing (AM) techniques have developed from just being employed in rapid prototyping methods to assist in manufacturing processes. The latter aims to produce finished parts that are economically feasible, with long-term stability, and with high properties. Moreover, AM processes do not require special or costly tooling for manufacturing the parts and have shown significant potential in reducing both time and cost of product development. As a new manufacturing technology, AM makes it possible to fabricate components with 3D-complex geometries that are hard or impossible to obtain using conventional manufacturing methods. Fused filament fabrication (FFF) is one of the most used AM technologies because of its fast production, cost-efficiency, ease of use, large material adaptation, and capability to produce complex components. This process is based on the deposit of liquefied thermoplastic polymer filament layer upon layer to fabricate 3D part from a 3D computer-aided design (CAD) system. In this process, a thermoplastic filament is extruded through a heated nozzle in a semisolid state, and it solidifies and bonds with the already extruded material.

Dimensional accuracy, quality, mechanical, and physical properties of a 3D part material in FFF process are dependent on various selected conflicting printing parameters. Hence, the study of these parameters is the most important in order to control the final performance and properties of manufactured parts. The conflicting parameters of fused deposition modeling (FDM) create much difficulty in determining appropriate parameters that will improve the final properties and part quality. As discussed, FFF

Industrial Strategies and Solutions for 3D Printing: Applications and Optimization, First Edition. Edited by Hamid Reza Vanaei, Sofiane Khelladi, and Abbas Tcharkhtchi.

process is one of the most complicated process, due to the large number of printing (controlling) parameters that you can select, with the goal of isolating which factors are active in the final AM products and to find an optimum combination of factor levels to accomplish a predetermined goal defined by the responses (part quality, dimensional accuracy, manufacturing time, mechanical properties, material consumption, etc…).

In this chapter, our objective is the understanding of the characteristics of FFF parts by studding the effect of the most processing parameters on the final properties of manufactured parts. The adopted approach is to make the link between the input factors and the macroscopic structural, thermal, and mechanical properties.

13.2 Controlling Factors of FFF Process

FFF is a complex process, where part quality and mechanical properties are affected by a large number of process parameters. The successful manufacturing of parts by FFF will, therefore, be achieved through optimized process factors aiming at both quality and reduced cost and production time. Figure 13.1 illustrates the cause-and-effect diagram of the adjustable FFF process factors. They can be classified under six categories, namely, 3D model, material, FFF machine, environmental factors, printing factors, and build direction. Some of these parameters, specifically the printing factors and environmental factors, can be easily adjusted in the control software to modify the final properties of the part.

Some main parameters are also described as follows:

- ✓ Layer thickness in 3D printing is defined as a measure of the layer height of each successive addition of material in the 3D-printing process in which layers are stacked. This printing factor is essentially the vertical resolution of the *Z*-axis. The number of layers needed to print a part determines the printing speed and thus the printing time required.
- ✓ Infill or layer angle is defined as the angle between the nozzle's path and the *X*-axis of the build table. The typical layer angles allowed are 0–90° or 0° to −90°.
- ✓ Infill density refers to the amount of material that occupies the internal part. It is usually defined as a percentage between 0 and 100, with 0% making a part hollow and 100%, completely solid. The infill pattern can be a grid, a triangle, or something more complex like concentric rings. Infill density can be adjusted depending on the function of the 3D-printed part. Visual parts and lightweight parts should have a low density, while functional parts should have a higher density.

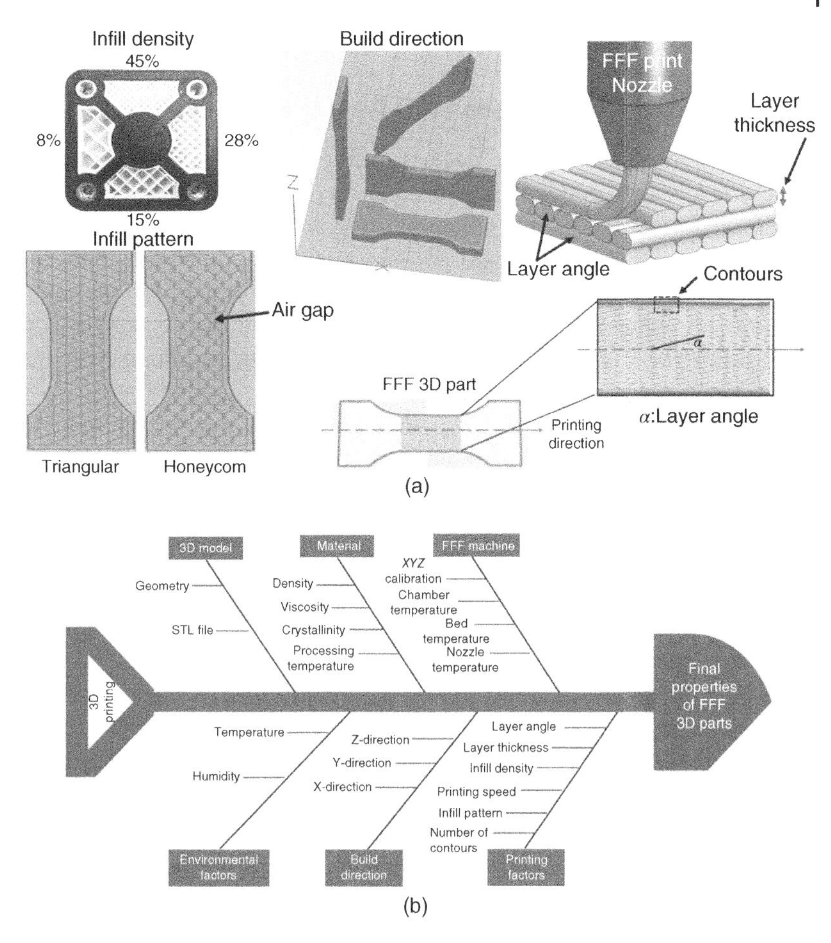

Figure 13.1 (a) Schematic of key parameters of FFF manufacturing process, (b) cause and effect diagram of FFF process parameters.

✓ The nozzle temperature defines temperature of the heating block, where the filament polymer achieves a viscous state. Nozzle becomes hot through heat conduction from heat source. Deposition strength of FFF type 3D-printed part is highly dependent on deposition temperature. The rheological performance of the deposited polymer and mechanical properties of the printed structure get affected mainly by the nozzle temperature during the FFF process [1].

✓ Print speed is the distance traveled by the extruder along the XY plane per unit time while extruding. Printing time depends on print speed, and the print speed is measured in mm/s.

13.3 Overview of Optimization

13.3.1 What Is "Optimization of 3D-Printing Parameters"?

To answer this question, we need to define what is optimization first. The simple way to define the optimization is that it is a tool, technic, and approach that can be used to minimize waste and maximize profit in order to have better efficiency. There are two types of optimization, namely, single-objective optimization (SOO) and multi-objective optimization (MOO), the difference is that in SOO only one factor can be studied at a time, unlike MOO multiple factors that can be taken in consideration at the same time.

Optimization generally involves determining the maximum or minimum value considering one or several objectives. In cases where several objectives are involved, the problem is known as a MOO. The objectives of any project or process are generated based on the problems or limitations associated with it. The approach of MOO is adopted over a single objective since in most real industrial processes, optimizing one aspect (factor) has a direct influence on the other parameters and, therefore, could cause conflicts among the optimized and other parameters.

In the FFF process, the objective is to achieve low surface roughness, high mechanical strength, low defect density (such as cracks and porosity), thermally stable prints, dimensionally accurate products, reduced printing time, and reduced material cost, just to mention a few. As such, just like other manufacturing processes, fused deposition modeling is a multi-objective problem and requires the MOO approach for optimal process and product quality. There are so many MOO methods utilized across various fields; some of them include genetic algorithms (GA), differential evolution (DE), pareto evolutionary algorithm (PEA), non-dominated sorting genetic algorithm-II (NSGA-II), particle swarm optimization (PSO), Hungarian algorithm (HA), analytic network process (ANP), and hybrid methods, among so many other techniques.

The most common method of optimization is the Taguchi methods, sometimes called robust design methods, developed by Genichi Taguchi to improve the quality of manufactured goods and more recently also applied to engineering, technologies, and AM (3D printing). Among all the Taguchi methods, in this chapter, we are going to focus more on the method called **response surface methodology (RSM)**.

RSM explores the relationships between several explanatory (dependent and independent) variables and one or more response variables. The main idea of RSM is to use a sequence of designed experiments to obtain an optimal response. Statistical approaches such as RSM can be employed to maximize the production of a special substance by optimization of operational factors. Of late, for formulation optimization, the RSM, using proper design of experiments (DoE), has become extensively used in all technology and industry fields such as 3D printing.

13.3.2 Response Surface Methodology (RSM)

RSM is an assortment of statistical and mathematical techniques with the only objective to analyze, using an empirical model, difficulties and problems given in a situation. Because of the complexity of some situations with a controllable variable that ends with no availability of the theoretical model, the empirical way is the best method or approach to get information about the relation between factors and responses, and this is the power of the RSM [2]. The objective of the RSM are the following:

- ✓ Estimating the experimental variability in reliable way.
- ✓ Guaranteeing the compatibility between the experimental data and the proposed approach/model.
- ✓ Providing the high level of efficiency with the respect to the constraints especially the economic cost, time, and other limitations.
- ✓ To make the decision-making possible under uncertain conditions, reducing the ambiguity.
- ✓ To generate knowledge in the experimental domain of interest.

The most extensive applications of RSM are in the industrial world, particularly in situations where several input variables potentially influence performance measures or quality characteristics of the product or process. These performance measures or quality characteristics are called the response. They are typically measured on a continuous scale, although attribute responses, ranks, and sensory responses are not unusual. Most real-world applications of RSM will involve more than one response. The input variables are sometimes called independent variables, and they are subject to the control of the engineer or scientist, at least for purposes of a test or an experiment.

In the literature, there are lots of experimental designs and approaches. Some of them are consequences of the optimality criteria and come from theoretical studies. Others have been generated to solve concrete problems. In addition, there are several computer packages that provide designs, based on an optimality criterion, just for the need of the user such as Minitab [2].

The relation between variables and response is theoretically described by a function that is the underlying physical mechanism of the problem under study. The existence of this relation makes the phenomenon under study sufficiently reproducible to be able to experiment with it and to extract conclusions:

$$y = f(U1, U2, \dots., Uk) + \varepsilon$$

ε represents other sources of variability that were not considered in f like the error in the determination of the response. The variables $U_1, U_2, \dots, U_k$ are the natural variables or factors. However, they are expressed in different units and vary in different ranges.

In general, the real function f is unknown and/or very complex, but it can be approximated by another simpler and easier polynomial function (first order or second order). The approximating model is based on the experimental data, and, as such, it is an empirical model whose finality is to adequately represent the response in the experimental domain [1].

$$y = \beta_0 + \beta_1 X_1 + \beta_2 X_2 + \cdots + \beta_k X_k + \varepsilon \tag{13.1}$$

$$y = \beta_0 + \sum_{i=1}^{k} \beta_i X_i + \sum_{i=1}^{k} \beta_{ii} X_{i^2} + \sum_{i<j} \sum \beta_{ij} X_i X_j + \varepsilon_i \tag{13.2}$$

β_i represents the expected change in response y per unit change in X_j when all the remaining independent variables are held constant.

Equation (13.1) is the proposed first-order response surface model that can describe the relationship between y and X_i.

Equation (13.2) is the proposed second-order model required in case where the curvature in the true response surface is strong enough to makes the first model inadequate.

13.3.3 Equation of Regression and ANOVA

Regression is a statistical tool to predict the dependent variable with the help of one or more than one independent variable. While running a regression analysis, the main purpose of the researcher is to find out the relationship between the dependent variable and the independent variable. In order to predict the dependent variable, one or multiple independent variables are chosen, which can help in predicting the dependent variable. It helps in the process of validating whether the predictor variables are good enough to help in predicting the dependent variable.

Regression analysis is the analysis of relationship between dependent and independent variable, as it depicts how dependent variable will change when one or more independent variable changes due to factors, formula for calculating it is $Y = a + bX + E$, where Y is dependent variable, X is independent variable, a is intercept, b is slope, and E is residual.

Analysis of Variance (ANOVA) is a statistical formula used to compare variances across the means (or average) of different groups. A range of scenarios is used to determine if there is any difference between the means of different groups. Some people question the need for ANOVA; after all, mean values can be assessed just by looking at them. But ANOVA does more than only compare means.

Even though the mean values of various groups appear to be different, this could be due to a sampling error rather than the effect of the independent variable on the dependent variable. If it is due to sampling error, the difference between the group means is meaningless. ANOVA helps to find out if the difference in the mean values is statistically significant. Typically, ANOVA is used in combination with other statistical methods that we are going to explain: **M**ain effects **P**lots–**P**areto chart–**O**ptimization chart.

13.3.4 Main Effect Diagram and Pareto Chart

The main effects plot (Figure 13.2a) is the simplest graphical tool to determine the relative impact of a variety of inputs (example: printing temperature, speed, and layer thickness) on the output of interest (example: elastic modulus). In analysis of variance, the main effects plot shows the mean outcome for each independent variable's value, combining the effects of the other variables. In other words, mean response values at each level of the process variable.

In the FFF processes the main effects plot is used for example to examine differences between level means for nozzle temperature, printing speed, and layer thickness factors. There is a main effect when different levels of a factor affect the (the tensile property, the yield stress, or the surface roughness) response differently. The importance of the factor can be judged by the curve variation and by the probability p-value that should be lower than the $alpha$ value set generally, which is equal to 0.05.

As shown in Figure 13.2a, the main effects plot was used to examine differences between level means for printing temperature, printing speed, and layer thickness controlling factors. Results showed that there is a main effect because different levels of all factors affect the elastic modulus response differently. The importance of the factor can be judged by the curve variation and the p-value recorded of the factor. A p-value upper than 0.05 with a horizontal curve implies a lesser impact of the input parameter on the measured response. The mean effect plot results (Figure 13.2a) show that printing at ~200 °C with a high printing speed (50 mm /s), and 0.2 mm layer thickness gives to *material* a rigid behavior with higher elastic modulus.

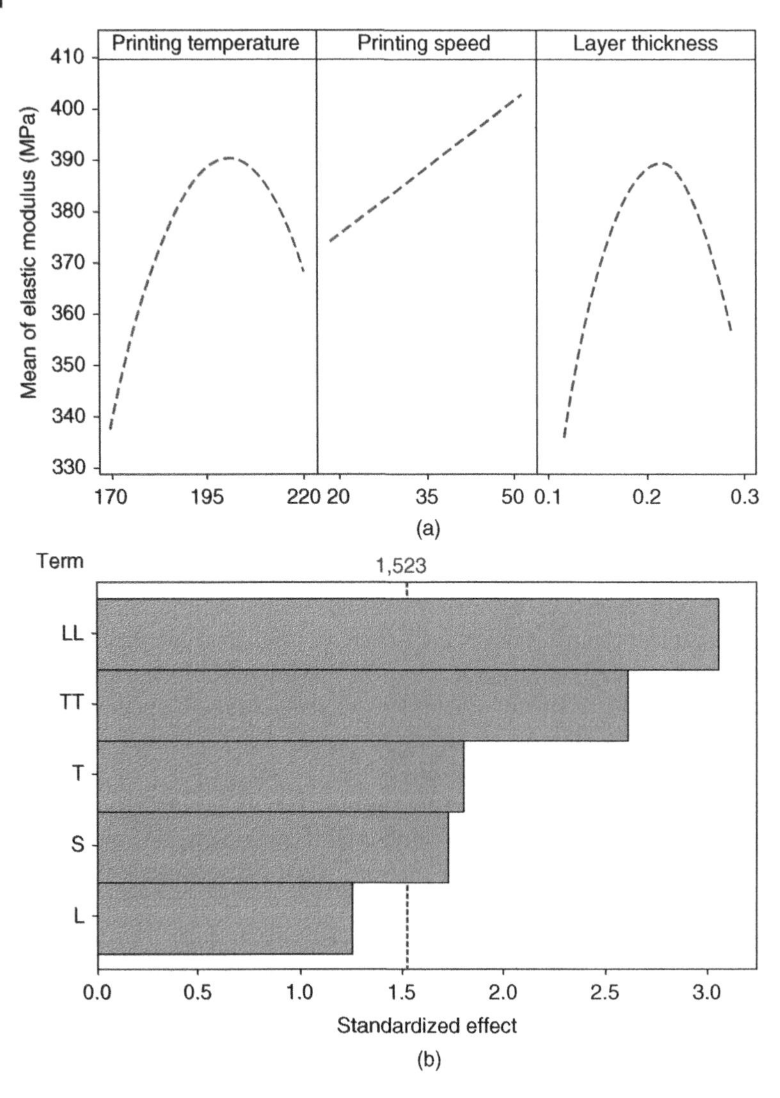

Figure 13.2 Example of (a) main effects plots and (b) Pareto chart.

In addition to main effects plots, Pareto charts are also used to rank from largest to smallest the effects of input parameters so that can be prioritized the improvement efforts regarding responses. These charts indicate the main and interaction effects generated by the ANOVA. They have also the advantage to identify the standardized effects of different linear, quadratic, and interacted terms of evaluated controlling parameters vs the corresponding standard value (1.523). Any effect beyond this standardized value line is statistically significant in the fitted model of the measured response.

We noticed from Figure 13.2b that all bars cross that reference line except (T) parameter, indicating that these parameters are statistically significant at the 0.05 level with their regression models terms. When evaluating the measured, the quadratic effect of L parameter is the most significant parameter, followed by the quadratic effect of T parameter. However, the linear effect of L input parameter is not statistically significant (p-value>0.05) in response model. One of Pareto charts' advantages is its ability to evaluate the interaction between effects. Any statistically significant interaction was observed, for example, in the measured response, meaning that the effect of input variables does not depend on each other.

13.3.5 Contour Plots, 3D Surface Plots, and Optimization Diagram

A contour plot (Figure 13.3a) is a graphical technique for representing a 3D surface by plotting constant z slices, called contours, on a 2D format. That is, given a value for z, lines are drawn for connecting the (x, y) coordinates where that z value occurs. The contour plot is an alternative to a 3D surface plot. We use a contour plot to explore the potential relationship between three variables. Contour plots display the 3D relationship in two dimensions, with X- and Y-factors (predictors) plotted on the x- and y-scales and response values represented by contours. A contour plot is like a topographical map in which x-, y-, and z-values are plotted instead of longitude, latitude, and elevation [3].

As discussed above, the RSM of DoE can provide a clear prediction as regards the significance of interactions and square terms of controlling factors. 3D surfaces produced by RSM can bring about visualization of the effect of factors in response to the entire range specified. A 3D surface plot (see Figure 13.2b) is a 3D graph that is useful for investigating desirable response values and operating conditions (factors).

A surface plot contains the following elements:

✓ Predictors on the x- and y-axes.
✓ A continuous surface that represents the response values on the z-axis.

After model's fitting of each response separately, determination of the predicted models and making the contour plots of the most influential parameters, it is necessary to optimize the responses and define the best combination of factors that maximize or minimize as much as possible

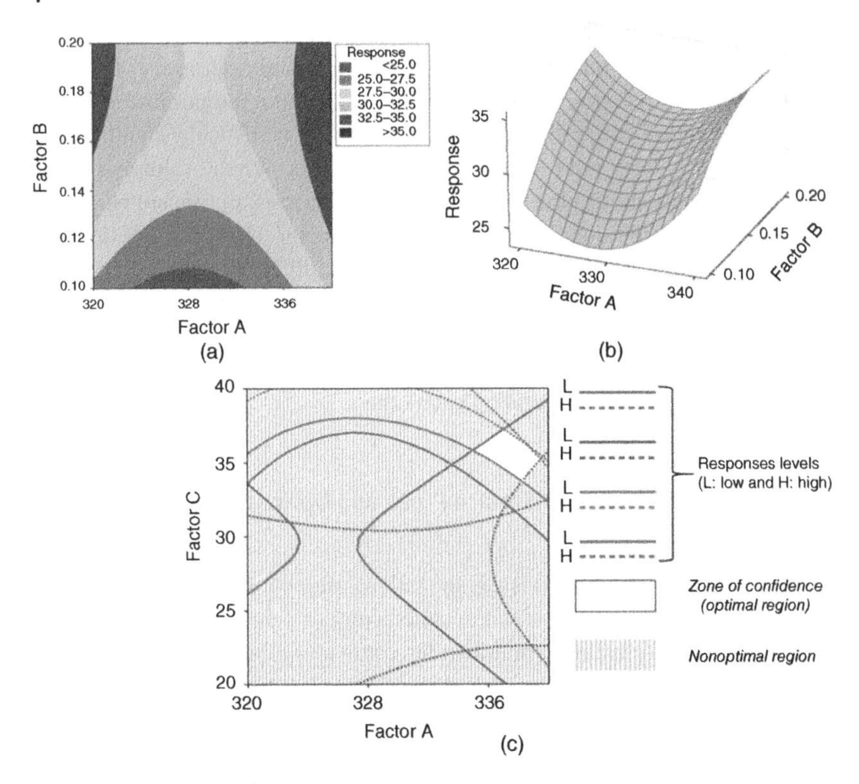

Figure 13.3 Example of 3D surface plot.

responses. The optimized factors will be combined together to identify a single response answering good response.

Overlaid contour plot is used to identify a feasible area where the predicted means of all response variables were in an acceptable range. As we can see in Figure 13.2c, the RSM optimization process helped to achieve the results of the interaction between two controlling factors (A and C), the shredded area is the nonoptimal region because it does not give an optimal characteristics, the non-shredded area is the optimal region where the test bar gave the optimal results in term of responses.

13.4 Advantages and Disadvantages of the Optimization

The fact that we can optimize 3D-printing parameters by controlling the factors is already an advantage, moreover, there is more advantages on many different sides.

On the design side, optimization of printing factors gives more freedom of design because sometimes a design cannot be achievable because of some factors that are not too efficient, after optimizing those factors and having control over the overall printing parameters, we could go further on the design of the 3D part, especially with the existence of the topological optimization that can be applied to minimize the weight of the part without losing any mechanical property. On the consumption side, optimizing the printing factors can help optimizing the power consumption because of the niched and homogeneous experiences that can be done with the interaction between several factors, so it gives less experiences to study without a lot of use of power, also the raw material cannot be ignored because it would have a better efficiency in term of raw material consumption. On the maintenance side, the optimization can extend the lifetime of the machine by balancing between the entries and the outcomes, also with the MOO, we have less experience and less part to print than SOO which helps the machine to work for longer time. Optimization of printing factors gives more probability to improve the quality of the 3D-printed parts in the overall perspective.

Also, it helps the user to minimize the chance of getting a failed 3D-printed part, and especially since we are studying the controlling factors and characteristics of 3D-printed parts using the FFF process because of its sensitivity to the outside noises, optimization can also include another aspect of the process monitoring that can detect the defect in printing and can correct them eventually. And finally, the optimization has a big bright side, it gives always the need and the curiosity to outperform any given task, problem, and difficulty. Also, with the tools offered these days, it can help the user to achieve more customized solutions without losing much time, power, or money and without losing the structural, morphological, or mechanical properties of the 3D-printed part by the FFF process.

The same as optimization has that much advantages, it has also some disadvantages or some limitations. First, the optimization of controlling factors of 3D-printed parts using FFF process needs to be continually updated because the outside factors or the noisy factors that are part of the environment can disturb the good printing process and cause defects either for the part or the machine itself. Second, with the exponential development in the 3D-printers market, there are always differences between machine and so one, it means that the optimization of factors which works perfectly in the printer A does not necessarily work in the printer B, and the difficulty is that it need to be adapted generally. Finally, another limitation of the optimization of the controlling factors is that it can be a little complicated in function of the number of factors that need to be controlled and the desired characteristics for the final 3D-printed part especially with the FFF process.

13.5 Optimization in 3D-Printing Perspective

For this part, we are going to see the different impacts of the optimization factors on the different aspects of 3D-printing FFF process. Starting with the first step before printing, which is the design, in this stage more the design is adapted to 3D printing more it will be efficient, that is why we always need to make parts with the design for additive manufacturing (DFAM). Moreover, the optimization has a direct impact on the printing time or the number of experiments needed for a study of some parameters, and it gives the chance to study the interaction between the parameters and more real results, and it affects directly the cost of the parts. The cost is always related to the printing time, and the quality of the part, so if optimization would help to optimize the parameters for better mechanical, structural, and morphological properties; therefore, it would minimize the cost of the parts 3D printed.

On the structural plan, the optimization of 3D-printing FFF process has a significant impact. For example, the optimization of the printing process parameters or factors may modify the esthetic finish of the 3D-printed parts, if the part needs to be less rough and glossier, we need to optimize the factors in order to get the best response objective which is surface roughness. Let us take the nozzle temperature as an example, during the extrusion process, the structural morphology of the polymer could change somehow, and it will affect the crystallinity which impacts the mechanical properties [4].

13.6 Optimization of 3D-Printing FFF Controlling Factors

In this section, the effect of the most important controlling factors was discussed in terms of their optimization and their impact on the final performance of the 3D-printed part. The selected physical factors are nozzle temperature, layer thickness, printing speed, and infill density. These factors have been selected for their great impact during the 3D-printing process and on the finale performances of manufactured parts.

13.6.1 Nozzle Temperature

As known, nozzle temperature has a great impact on the mechanical, structural and thermal properties of 3D fused filament fabricated parts because of its impact on the crystallinity, the layer adhesion and also on the morphological behavior of the parts. In this part, we are going to focus on the nozzle temperature parameter also called extrusion temperature, where we are going to oversee the most important studies in this field.

Deng and Zeng [5] did a research about the optimization of mechanical properties of the poly-ether-ether-ketone (PEEK) using the FFF process, there purpose is to investigate the influence of the printing temperature, printing speed, layer thickness, and filling ratio on the mechanical properties of the PEEK samples. They have used an L9 orthogonal design (with 9 runs or experiments) for the four selected factors. Optimization of results showed that printing speed of 60 mm/s, layer thickness of 0.20 mm, printing temperature of 370 °C, and a filling ratio of 40% exhibit printed PEEK's parts' highest elastic modulus, tensile strength, and elongation. They conducted that the printing temperature also significantly affects the level of crystallinity and, thus, the mechanical properties. Too high extrusion temperature may cause the material degradation, which cannot retain its shape upon deposition, resulting in filament deformation and dimensional inaccuracy. However, if the extrusion temperature is not high enough, the material does not have enough thermal energy and time to fully melt, which could result in the low layer adhesion and nozzle clogging.

In another study done by Alafaghani et al. [4], where they did an experimental optimization of the FFF process parameters. They started with setting up an experimental plan with the values of the six processing parameters, namely, build direction, infill percent, print speed, extrusion temperature, layer thickness, and infill pattern and they got 18 runs to test. After printing, they did a dimensional accuracy test to see how can a parameter influence the dimensions of the parts, and they found out that the extrusion temperature has a significant effect on the accuracy, as the extrusion temperature increases the error increases due to the increased flow rate because it the filament become more viscous which allows more raw material to be deposit. After the tensile test on the specimens, they found that the mechanical properties depend on the extrusion temperature to a certain limit, where it starts dropping. To conclude, this article shows the importance and the impact of the nozzle temperature on the dimensional accuracy and the mechanical properties.

Francisco and coworkers [6] studied the influence of the 3D-printing parameters on the mechanical properties of polylactic acid also known as PLA, they aimed to determine the influence of 3D-printing parameters such as infill density, extrusion temperature, raster angle, and layer thickness on the mechanical properties. With the help of a DoE, they established a table with all the possibilities of interactions between the selected parameters, and they ended with 24 experiments. ANOVA results confirmed that the adhesion between layers improves when there is a higher nozzle temperature (220 °C), but it tends to become more fragile. Moreover, at a lower temperature (200 °C) the adhesion drops and that makes the part more ductile with high elongation at break and toughness.

13.6.2 Layer Thickness

As mentioned before, the layer thickness parameter is an important parameter that has many impacts on the surface roughness of the 3D-printed parts, on the time/cost of the part, and mostly on the mechanical properties because the bigger the layer thickness is the less is the adhesion that means more voids that can be framed as porosity and by extension can be the source of cracks.

Anitha et al. [7] investigated the effects of some important FDM process parameters on surface roughness of acrylonitrile butadiene styrene (ABS) prototype. The Taguchi's design matrix, signal-to-noise ratio (S/N), and ANOVA were used in this study. Three process parameters, including layer thickness, road width, and speed of deposition, were considered. This study revealed that the factor having the most important influence on the surface roughness was the layer thickness because what makes a surface more or less rough is the resolution which depends on the air gap between the printed layers, more the layer thickness is high less the surface roughness, compared to road width and speed. It was also revealed that there was inverse relation between layer thickness and surface roughness.

Another study was done by Nancharaiah et al. [8] who also applied Taguchi method and ANOVA technique to identify the key factors that influenced the dimensional accuracy of 3D FFF-printed ABS parts. The input variables such as layer thickness, road width, raster angle, and air gap were considered. They concluded that layer thickness and air gap significantly affected the accuracy of FDM parts. However, in this study, optimum settings of layer thickness, road width, raster angle, and air gap in the range were not addressed [8].

El Magri et al. [9] investigated the effects of some important FDM process parameters on the tensile behavior of 3D-printed acrylonitrile styrene acrylate (ASA) material prototype in Z direction. The RSM, ANOVA, and central composite design (CCD) were used in this study. Three process parameters including layer thickness, nozzle temperature, and printing speed were considered. This study reveals that the layer thickness parameter is the dominant parameter of tensile properties through the Z direction because the bigger the layer thickness is, the less is the adhesion that means more voids that can be framed as porosity and by extension can be the source of cracks, layer thickness also appear to affect strongly the strain at break. Higher elongation values were recorded for samples produced at the highest extrusion temperature and layer thicknesses. These results can be explained by better diffusion of polymer chains between the freshly printed and the previously printed layer. After the optimization using RSM, they found the optimum values of the parameters under certain criteria with a layer thickness of 0.155 mm.

13.6.3 Printing Speed

Deng et al. [5] investigated the mechanical properties optimization of PEEK by FFF process. The L9 orthogonal and range analysis methods were used in this study. Four process parameters such as layer thickness, printing temperature, infill density, and printing speed were considered. This study revealed that the range analysis showed the optimal behavior of tensile strength comes with a printing speed of 60 mm/s along with the other parameters. From another response behavior of elongation rate, the optimum printing speed comes with a value of 20 mm/s along with the other parameters. As for the elastic modulus, the optimal behavior of elastic modulus comes with a printing velocity value of 60 mm/s. With the use of comprehensive analysis, they ended up having 60 mm/s as a printing speed for optimal results. From this study, we can see the impact of printing speed on the mechanical properties, it is an important factor because it powers controlling the crystallinity, the thickness, the temperature diffusion, the time, and the cost. For the speed, it is always a compromise especially with printing temperature.

Attoye et al. [10] investigated the correlation between process parameters and mechanical properties of 3D FFF-printed PLA and ABS parts. DoE is used to determine optimized values of the process parameters for each type of material. ANOVA was used to compare the significance of the effect of processing parameters on the mechanical properties. The study revealed that the effect of the printing speed on the build time was not significant. However, the mechanical analysis shows a more discernable influence of the printing speed parameter on the build process and final product mechanical properties.

Alafaghani et al. [4] did a study to examine the influence of FDM processing parameters on the final parts characteristics. A four-level scale method was used to investigate each one of these processing parameters building direction, infill percent, print speed, extrusion temperature, layer height, and infill pattern. In this study, they reported that the dimensional accuracy is affected by many parameters among them speed printing. It was shown that the mechanical properties are basically affected by many parameters especially the printing speed because that speed can give time to form the crystals or to form monocrystalline material for better mechanical resistance and behavior, or it can be too fast to let the material relaxes, and in this case, it will not have the optimal mechanical properties. Sometimes, an annealing treatment or heating treatment is done to give the crystals the chance to be reborn.

In other work, it was found that printing speed is the second most important condoling parameter because it affects significantly tensile strength and Young's modulus of printed ASA's samples. Authors found that printing speed mainly influences the interlayer bonding and the welding quality between layers. Because there is always a compromise between printing speed and the printing temperature, and we just have to find the balance.

13.6.4 Infill Density

3D-printed parts are typically not produced with a solid interior. Instead, the printing process uses different pattern types for interior surfaces. This greatly reduces cost due to reduced material usage and print time, while moderately reducing strength. The density of this pattern is referred to as the infill percentage.

This printing parameter significantly impacts the ultimate tensile strength [11]. It was confirmed that the infill percentage is proportional to the strength of the material [12]. In fact, infill density was found to influence the surface finish and dimensional accuracy of the parts. The best of these two properties is obtained with high infill density [13]. The same was observed by Camargo et al. [14], as it was concluded that increasing the infill density exhibits an enhancement in the mechanical properties except for impact energy that showed a decrease as infill increases. Similar results were revealed in a study, where the mechanical properties of PLA-graphene were evaluated by varying the infill density from 0 to 100% [15]. For ASA material, for example, it was concluded that the higher is the infill density, the higher is the tensile strength along the print direction [16]. In other study using the RSM, it was revealed that infill density and layer thickness are most significant factors on tensile strength and hardness [17]. In fact, changes in the infill density exhibit a variation in the meso and macro structures of 3D-printed parts [13]. Generally, achieving higher ultimate tensile strength with 100% infill density is explained by the fact that at this infill percentage, a higher surface area of bonding between each layer is achieved [11]. Even for the flexural strength and compressive strength, it was observed that they are both proportional to infill percentage [12].

Gabriel et al. [18] evaluated the infill effect on mechanical properties of consumer 3D-printing materials, namely, PLA, ABS, polyethylene terephthalate glycol (PETG), various nylons, polycarbonate (PC)/Acrylonitrile Butadiene Styrene (ABS), and ASA. They concluded that for all the materials, the infill percentage affect significantly the modulus, elongation, and failure mode. Always about the mechanical properties, the FDM parameter does influence the stiffness as well as the impact strength [19].

References

1 Elhattab, K., Bhaduri, S.B., and Sikder, P. (2022). Influence of fused deposition modelling nozzle temperature on the rheology and mechanical properties of 3D printed β-tricalcium phosphate (TCP)/polylactic acid (PLA) composite. *Polymers (Basel).* 14 (6): https://doi.org/10.3390/POLYM14061222.

2 Khuri, A.I. and Mukhopadhyay, S. (2010). Response surface methodology. *Wiley Interdisciplinary Reviews: Computational Statistics* 2 (2): 128–149. https://doi.org/10.1002/WICS.73.

3 R.H. Myers, Douglas C. Montgomery, Christine M. Anderson-Cook, Response Surface Methodology: Process and Product Optimization Using. p. 894, 2016. https://www.wiley.com/en-ca/Response+Surface+Methodology%3A+Process+and+Product+Optimization+Using+Designed+Experiments%2C+4th+Edition-p-9781118916018 (accessed July 25, 2022).

4 Alafaghani, A., Qattawi, A., Alrawi, B., and Guzman, A. (2017). Experimental optimization of fused deposition modelling processing parameters: a design-for-manufacturing approach. *Procedia Manufacturing* 10: 791–803. https://doi.org/10.1016/J.PROMFG.2017.07.079.

5 Deng, X., Zeng, Z., Peng, B. et al. (2018). Mechanical properties optimization of poly-ether-ether-ketone via fused deposition modeling. *Materials (Basel)* https://doi.org/10.3390/ma11020216.

6 Tor, S.B., Francisco, J., and Fernandes, M. Study of the influence of 3D printing parameters on the mechanical properties of PLA. Proceedings of the 3rd International Conference on Progress in Additive Manufacturing (Pro-AM 2018), 547–552. https://doi.org/10.25341/D4988C.

7 Anitha, R., Arunachalam, S., and Radhakrishnan, P. (2001). Critical parameters influencing the quality of prototypes in fused deposition modelling. *Journal of Materials Processing Technology* 118 (1–3): 385–388. https://doi.org/10.1016/S0924-0136(01)00980-3.

8 Nancharaiah, T., Ranga Raju, D., and Ramachandra Raju, V. (2010). An experimental investigation on surface quality and dimensional accuracy of FDM components. *International Journal on Emerging Technologies* 1: 106–111.

9 El Magri, A., Ouassil, S.E., and Vaudreuil, S. (2022). Effects of printing parameters on the tensile behavior of 3D-printed acrylonitrile styrene acrylate (ASA) material in Z direction. *Polymer Engineering and Science* 62 (3): 848–860. https://doi.org/10.1002/PEN.25891.

10 Attoye, S., Malekipour, E., and El-Mounayri, H. (2019). Correlation between process parameters and mechanical properties in parts printed by the fused deposition modeling process. *Conference Proceedings of the Society for Experimental Mechanics Series* 35–41. https://doi.org/10.1007/978-3-319-95083-9_8.

11 Jackson, B., Fouladi, K., and Eslami, B. (2022). Multi-parameter optimization of 3D printing condition for enhanced quality and strength. *Polymers (Basel).* 14 (8): https://doi.org/10.3390/polym14081586.

12 Quader Shurjeel, A., Pothula, N., and Punna, E. (2021). Experimental investigation of strength properties of 3D printed ABS composites. *E3S Web of Conferences* 309: 01148. https://doi.org/10.1051/e3sconf/202130901148.

13 Abeykoon, C., Sri-Amphorn, P., and Fernando, A. (2020). Optimization of fused deposition modeling parameters for improved PLA and ABS 3D printed structures. *International Journal of Lightweight Materials and Manufacture* https://doi.org/10.1016/j.ijlmm.2020.03.003.

14 Camargo, J.C., Machado, Á.R., Almeida, E.C., and Silva, E.F.M.S. (2019). Mechanical properties of PLA-graphene filament for FDM 3D printing. *International Journal of Advanced Manufacturing Technology* 103 (5–8): 2423–2443. https://doi.org/10.1007/s00170-019-03532-5.

15 Alvarez, K.L., Lagos, R.F., and Aizpun, M. (2016). Investigating the influence of infill percentage on the mechanical properties of fused deposition modelled ABS parts. *Ingeniería e Investigación* 36 (3): 110–116. https://doi.org/10.15446/ing.investig.v36n3.56610.

16 Raam Kumar, S., Sridhar, S., Venkatraman, R., and Venkatesan, M. (2020). Polymer additive manufacturing of ASA structure: influence of printing parameters on mechanical properties. *Materials Today Proceedings* 39 (xxxx): 1316–1319. https://doi.org/10.1016/j.matpr.2020.04.500.

17 Srinivasan, R., Pridhar, T., Ramprasath, L.S. et al. (2020). Prediction of tensile strength in FDM printed ABS parts using response surface methodology (RSM). *Materials Today Proceedings* 27 (xxxx): 1827–1832. https://doi.org/10.1016/j.matpr.2020.03.788.

18 Johnson, G.A. and French, J.J. (2018). Evaluation of infill effect on mechanical properties of consumer 3D printing materials. *Advanced Technology Innovation* 3 (4): 179–184.

19 Popescu, D., Zapciu, A., Amza, C. et al. (2018). FDM process parameters influence over the mechanical properties of polymer specimens: a review. *Polymer Testing* https://doi.org/10.1016/j.polymertesting.2018.05.020.

14

Machine Learning in 3D Printing

Mohammadali Rastak[1], Saeedeh Vanaei[2], Shohreh Vanaei[3], and Mohammad Moezzibadi[4]

[1] *Department of Mechanical, Industrial and Aerospace Engineering, Concordia University, Montreal, Quebec, Canada*
[2] *Department of Mechanical, Industrial and Manufacturing Engineering, University of Toledo, Toledo, OH, 43606, USA*
[3] *Department of Bioengineering, Northeastern University, Boston, MA, USA*
[4] *Arts et Metiers Institute of Technology, CNAM, LIFSE, HESAM University, 75013 Paris, France*

14.1 Introduction

In the modern era, intricate components are in tremendous demand for a variety of applications, including architecture, industry, medical care, and scientific research. Extensive research and analysis have been conducted on the manufacturing process in an effort to simultaneously improve quality and performance while reducing cost and production time. In this regard, additive manufacturing (AM) and 3D printing are among the techniques that will replace many conventional manufacturing processes within the next few years.

Numerous industries have utilized 3D printing to fabricate a vast array of structures. The ability to manufacture complex geometries is one of the primary advantages of this technique, and the quality of the final product is likely preferred to the majority of other manufacturing methods. Nevertheless, there are numerous issues that must be addressed. Poor fusion, for instance, results in porosity and wrapping of the component due to internal residual stress [1]. In addition, some physical parameters, such as printing speed and layer thickness, contribute to the inconsistency of product quality, which is the greatest obstacle in additive manufacturing and 3D printing [2].

These considerations have substantial effects on the final product's quality. In addition, finding answers to these problems might not be straightforward owing to the fact that 3D printing is a complicated process

Industrial Strategies and Solutions for 3D Printing: Applications and Optimization, First Edition. Edited by Hamid Reza Vanaei, Sofiane Khelladi, and Abbas Tcharkhtchi.

that calls for a significant amount of effort and takes a lot of time before it can be optimized. Moreover, individual study in this field is inadequate. In other words, a good grasp of several parameters and profoundly studying interdisciplinary fields are necessary, such as solid and liquid interaction, thermal and mechanical interaction, and fluid dynamics, to name a few [1].

In recent years, the field of artificial intelligence (AI) has seen the rise of a robust subset known as machine learning (ML). It is possible to examine the relationship between the parameters of a physical problem by using ML in order to make a prediction about the effectiveness of the labels or outputs. To put it another way, ML can be used as a tool for optimizing difficult problems by employing the utilization of training data. Among the many examples that contributed to ML's rise to prominence are medical diagnostic, smart manufacturing, and material properties [3]. Therefore, machine learning has the potential to apply algorithms in 3D printing, thereby improving the performance of the systems and revealing the systems of the next generation of 3D printing. Practitioners of additive manufacturing can benefit from machine learning algorithms in a number of ways, including a reduction in overall costs, an improvement in product quality, and an optimization of the fabrication process [2]. Excitingly, AI, and particularly ML, can reach the goal tolerances of the part by detecting the deviation in the geometry during the manufacturing process and compensating for the defect after the fact. This is a huge step forward in the field of artificial intelligence. Fundamental of AI/ML are introduced in this chapter. Furthermore, the potential of AI/ML in 3D printing (Section 14.4.1) and how it can be applied in various aspects (Sections 14.4.2 and 14.4.3) such as defect detection, process optimization, cost estimation, and dimensional accuracy are discussed in the following sections.

14.2 Literature Review

Using the ML technique, Zhu et al. [4] attempted to mitigate shape deviation problems and improve tolerancing issues in AM. They used Bayesian inference to determine the deviation patterns, and the resulting accuracy is excellent. Tapia et al. [5] used the Gaussian model to predict the depth of the melt pool in the metal-based laser powder bed fusion process. In order to accomplish this, the ML model considers laser power, scan speed, and laser beam size as key features. In another study, Tapia et al. [6] introduced a Gaussian process model to estimate porosity using the selective laser melting (SLM) technique. In the first step, a Gaussian process was developed to model porosity as a function of the SLM parameters, followed by the application of Bayesian inference to estimate the parameters and porosity. In order to create high-density parts for the SLM process, Kamath [7] combined simulations and experiments. It has been demonstrated that the methodology can be used as a supplement to the physical model and that it can aid even more sophisticated processes. Meng and Zhang [8] created a Gaussian process model to predict the remelted depth based on the combined laser power and laser scan speed.

Caiazzo and Caggiano [9] implemented convolutional neural networks (CNN) to determine the relationship between geometrical parameters as output and laser metal deposition parameters and demonstrated that deep learning (neural network) has the capability to predict these parameters, resulting in a specified geometry. Zhang et al. [10] implemented a deep learning network in order to predict the tensile strength of FDM-manufactured parts' components. In this study, a long short-term memory network was used to characterize some parameters of the printing layers, and a layer-wise relevance propagation was executed, resulting in a useful method for perdition. Li et al. [11] attempted to use the ML approach and offline training and online prediction to predict the surface roughness of fused filament fabrication-manufactured parts. Mozaffar et al. [12] introduced the gated recurrent unit formulation in the recurrent neural network (RNN) for predicting the thermal history in the directed energy deposition (DED) technique. In the research carried out by Song et al. [13], the technique of Support Vector Regression was utilized to make predictions regarding the elemental makeup of the laser additive manufacturing process. In addition, it has been demonstrated that the method achieves the best results when contrasted with other approaches such as the calibration curve method, partial least square, and even artificial neural networks.

In their study on liquid metal jet printing, Wang et al. [14] proposed a solution to the issue of an uncertain influence-based process. This would allow the process to function more reliably. In order to accomplish this, they implemented a closed-loop control framework as well as a neural network in order to stabilize and improve the printing process. Hierarchical clustering of the AM was made possible thanks to the work of Yao et al. [15]. They demonstrated the potential for inexperienced designers to use this approach by applying the suggested method to the components of an R/C car. In a different piece of research carried out by Zhang et al. [16], a vision system was used in conjunction with support vector machines (SVM) and CNN to monitor the quality of the individual parts. The findings indicate that the classification accuracy can be improved to an accuracy of 90.1% by utilizing a combination of features. In conclusion, it has been demonstrated that CNN is superior to SVM in terms of performance when it comes to quality identification in the powder-bed fusion technique. A real-time control system for the LPBF machine was suggested by Scime and Beuth [17], who proposed using in situ monitoring instead. In this study, the algorithm implements an unsupervised machine learning algorithm, which enables it to recognize and classify anomalies. Khanzadeh et al. [18] focused their attention on the relationship between defects and a few characteristics of the melt poll found in the AM. As supervised learning methods, they experimented with a variety of classification algorithms, including decision tree (DT), K-nearest neighbor (KNN), and SVM. According to their findings, KNN has the highest level of accuracy when compared to other classification approaches for the classification of melt pools, whereas decision tree has the lowest level of error rate when it comes to the incorrect identification of melt pools (pores).

In a separate piece of research carried out by Samie Tootooni et al. [19], multiple classification algorithms such as decision tree, KNN, neural network, naive Bayes, SVM, and sparse representation were utilized in order to classify the AM parts in accordance with the dimensional variation. The SVM approach was used by Aoyagi et al. [20], who wanted to monitor the surface quality and predict whether the process condition would fall into the good or bad categories. When utilizing this technique, one of the benefits is that the number of experiments needed to obtain the optimal process can be lowered. This is one of the advantages. Ye et al. [21] presented a novel method using deep belief networks (DBN) and acoustic signals to address the issue of defect detection in SLM. By using feature extraction and signal preprocessing, DBN approach worked properly, and the results show its superiority among other methods, like SVM and multi-layer perceptron (MLP). Shen et al. [22] utilized a machine learning approach, more specifically 3D deep learning, in order to solve the problem of low accuracy of the parts produced by 3D printing. They proposed a framework for calculating the inverse function network, which would ultimately result in error compensation and an improvement in accuracy. The framework for cost-driven decision making that Jafari-Marandi et al. [23] presented uses in situ melt pool images. They assert that the theoretical foundation of this methodology not only has the ability to develop an in-process control approach, but also has the potential to improve the mechanical properties of the components.

Gobert et al. [24] developed an in situ defect detection system for powder bed fusion by using a supervised machine learning technique to implement an SVM model. The accuracy of this in situ classifier is greater than 80%, according to the results of the cross-validation. In addition, they reached the conclusion that it is very useful and enables one to make an informed decision to take several images under a variety of lighting conditions.

Table 14.1 ML applications in the 3D printing research.

AM processes	ML models	Inputs	Outputs	References
FDM	GP	Geometric parameters	Deviation	[4]
L-PBF	GP	Scan speed and laser parameters	Melt pool depth	[5]
SLM	GP	Scanning speed and laser power	Porosity	[6]
SLM	GP, RT	Laser power and speed	Melt pool depth	[7]
L-PBF	GP	Laser power and speed	remelted depth	[8]
L-DMD	ANN	Scanning speed, laser power, etc.	Geometrical parameters	[9]
FDM	SVM, RF	Temperature, printing speed, etc.	Tensile strength	[10]
FFF	SVR, RR, etc.	Print speed, thickness, etc.	Surface roughness	[11]
DED	RNN	Deposition time, toolpath feature, etc.	Thermal history	[12]
LIBS	SVR	Line intensity ratio, integrated intensity	In situ composition	[13]
LMJP	NN	Droplet features	Voltage	[14]
SLM	SVM	Target components	AM features	[15]
L-PBF	SVM, CNN	Melt pool, plume, spatters	3 classes	[16]
L-PBF	Bag-of-words	Images	Anomaly detection	[17]
DLD	DT, KNN, etc.	Melt pool characteristics	Porosity	[18]
FFF	NB, SVM, etc.	Dimensional variations	Infill	[19]
PBF	SVM	Energy density,surface parameter, etc.	Surface quality	[20]
SLM	DBN, MLP, SVM.	Acoustic signals	Defect detection	[21]
...	CNN	3D voxel grid	Deformation	[22]
Laser-AM	KNN, NN, etc.	Thermal features	Porosity	[23]
PBF	SVM	CT images	Defect detection	[24]
SLM	CNN	Images	Fault detection	[25]
SLM	SCNN	AE signals	Quality	[26]

Angelone et al. [25] present a novel bio-intelligent AM system that applies CNN. This system has been considered for the purpose of fault detection on SLM. Sectioning the specimens can be helpful in supporting the initial defect classification, which has a significant impact on the performance of the CNN, which is one of the main conclusions that they came to. Shevchik et al. [26] used spectral CNNs and acoustic emission features for the in situ quality monitoring of AM in order to investigate the product's quality. This was done so that they could determine how good the product was.

All in all, ML applications with name of the processes, ML algorithms, input parameters and outputs for some recent studies are listed in Table 14.1.

14.3 3D Printing: Applications and Obstacles

3D printing process is a type of digital manufacturing that builds a structure, typically complex shapes, by stacking layers in 3D. This technology enables the production of a vast array of components with complex shapes. This technology meets the needs of numerous global applications and industries, including the automotive, aerospace, food, fashion, medical, architectural, electric, and electronic sectors [27]. Other notable advantages of 3D printing include design flexibility, waste reduction, mass customization, and rapid prototyping, in addition to the ability to produce complex shapes [28]. In contrast, there are a number of disadvantages associated with 3D printing, such as time consumption, geometrical aspects, and material requirements, that may restrict the application of this manufacturing technology [29]. Moreover, as the volume of data in a manufacturing environment grows, new big data issues arise, necessitating the implementation of an integrated data infrastructure [30].

The application spectrum of 3D printing is rapidly expanding, encompassing diverse industries ranging from healthcare and aerospace to consumer goods and architecture. In the medical field, 3D printing is revolutionizing patient-specific care, enabling the creation of customized implants, prosthetics, and even intricate anatomical models for surgical planning. Aerospace engineers leverage 3D printing to fabricate lightweight yet robust components, optimizing fuel efficiency and performance. In the realm of consumer goods, this technology empowers designers to craft innovative and personalized products, while architects explore its potential to rapidly prototype intricate architectural models. The automotive sector embraces 3D printing for producing complex prototypes, specialized parts, and even entire vehicles, ushering in a new era of agility and customization.

Despite its remarkable potential, the 3D printing process confronts several challenges that must be navigated to fully unlock its benefits. Material limitations stand as a significant obstacle; while an increasing array of materials can be used in 3D printing, finding the right material with the required properties for a specific application remains a challenge. Achieving consistent quality and accuracy across large-scale production runs is another hurdle, often due to variations in the printing process and the need for meticulous calibration. Design complexity poses challenges too, as creating intricate structures can lead to issues like overhangs that require supports and subsequent post-processing. Additionally, post-processing itself can be cumbersome, demanding labor-intensive steps to refine surface finishes, remove support structures, and ensure part integrity. The relatively slow speed of 3D printing, especially for larger objects, compared to traditional manufacturing methods is also a concern in industries where time-sensitive production is crucial. Finally, the cost of 3D printing equipment, materials, and maintenance can be a barrier for small businesses and startups seeking to adopt this technology.

As a result, while 3D printing holds incredible promise across various industries, it is essential to recognize and address the obstacles that can hinder its seamless integration and widespread adoption. Ongoing research, technological innovation, and collaboration among stakeholders are vital to overcome these challenges and further propel the evolution of 3D printing.

14.4 AI/ML and 3D Printing

14.4.1 Role of AI/ML in 3D Printing

Recent applications of machine learning include engineering, medical applications, and the Internet of things. In this regard, AI, particularly ML, can significantly improve the performance of additive manufacturing and broaden the applicability of 3D printing. Intriguingly, 3D printing combined with AI can be a smart manufacturing technology that detects errors and resolves problems. Under these conditions, human participation is diminished or even unnecessary.

Four types are commonly used to classify ML algorithms for use in the AM: supervised, unsupervised, semi-supervised, and reinforcement learning. In supervised learning, algorithms use the training set to learn from a set of known examples in order to predict as precisely as possible the unknown output in the test set [31]. As depicted in Figure 14.1, each data (with features $X_1, X_2, \dots, X_n$) corresponds to an output Y (supervised

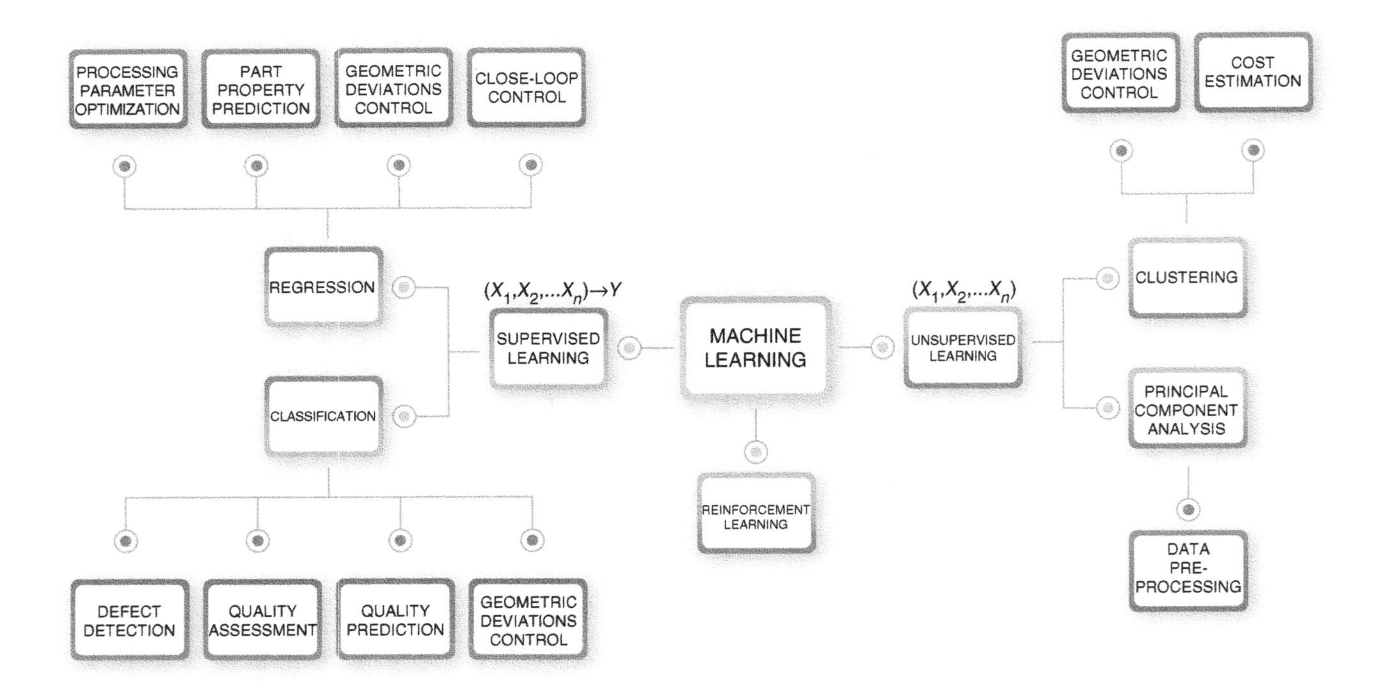

Figure 14.1 Different type of machine learning categories based on the dataset [2].

learning). It is important to note that the training set may contain a large number of instances. In unsupervised learning, the outputs are not given, so algorithms must find a good way to cluster the data, such as by grouping web pages by their subject [32]. Keeping this in mind, the similarity of the instances based on their features ($X_1, X_2, \ldots, X_n$ in Figure 14.1) should be considered when clustering them. After clustering the data, it is also possible to assign some labels to each cluster. In some references, another category known as semi-supervised learning is also mentioned. This category falls somewhere between the two ML groups previously mentioned. In other words, when working with a large amount of data, labeling can become a time-consuming and expensive process, so semi-supervised learning can be utilized. The last group is reinforcement learning. In this situation, it is assumed that training data can only serve as a rough indicator of whether a given action is correct [33]. Figure 14.1 depicts several real-world applications of machine learning in 3D printing [2].

Due to the fact that parameter estimation typically requires it, supervised learning can be used in the majority of manufacturing applications. Based on the type of output, there are two categories of supervised learning. A regression can be chosen if the label is a continuous parameter such as temperature or mechanical property. Alternatively, classification must be considered when categorical output is available, such as defective or not.

In additive manufacturing, the functionality of ML regression algorithms includes the prediction of property, optimization of manufacturing process parameters, geometric deviation control, and also cost prediction [2]. ML regression models in additive Manufacturing primarily employ artificial neural networks and Gaussian processes in this regard [2]. It is worthwhile to mention that multi-layer perceptron, or MLP, is typically used to refer to artificial neural networks or ANN.

It is notable to mention that despite the same application, ML models are in conjunction with the type of the data, such as defect detection, which can be obtained using supervised and unsupervised learning [2]. It is notable to mention that having adequate information of data is a key element for applying ML in any field, particularly in 3D printing. Furthermore, cleaning and organizing the datasets and finding the dimension of the datasets, which depends on the number of instances and features, should be taken into account before applying ML. Moreover, some features might not be as important as others. It is worthwhile to indicate that no algorithm is the best choice among other algorithms in ML, and each one has strengths and weaknesses. Hence, choosing an appropriate algorithm is an important task that is in conjunction with the available data. In other words, ML cannot be practical and capable when there is a limitation of available data. With this in mind, a plethora of data that is not biased toward any labels and reflects a real sample is necessary when any ML algorithms need to be trained. While NN can be appropriate for a complicated problem, GP might be a better choice when the degree of complexity is low [2]. The classification can be used if the outputs are the parameters, such as having a defect or good quality (as binary problems) or any categorical applications with more than two classes (multiclass problems). In this case, decision tree, SVM, and CNN are potential algorithms in ML.

14.4.2 ML Algorithms Review

It should come as no surprise that ML has emerged as one of the most significant subfields of both information technology and artificial intelligence in recent years [34]. ML refers to the collection of computational algorithms and models that attempt to simulate human intelligence [35]. In this particular scenario, many different classification strategies in ML have been suggested [36].

The KNN and DT machine learning algorithms are two popular approaches that can be used to perform on both categorical and numerical data. KNN makes an attempt to predict the outcome of new data by comparing it to all of the available or training data points, then comparing those similarities or distances to the new data. KNN is a nonparametric learning algorithm because it does not consider any assumptions about the dataset [37]. This makes KNN a nonparametric learning algorithm. In addition to this, KNN is a lazy learner [38]. On the other hand, DT utilizes a series of binary selections to partition the data into regions that are aligned along an axis. Both methods are still utilized in various machine learning applications.

The primary purpose of some algorithms is to identify certain visual contents within a picture or a texture. For example, the classification of textual data can typically be accomplished by utilizing either a generative classification (such as naive Bayes) or a discriminative classification (such as logistic regression), or both of these classification approaches. Text classification enables the categorization of texts written in natural languages according to the categories that are the most appropriate fit for the texts' individual contents [39]. In this context, certain words need to be taken into consideration in order to play the part of distinguishing characteristics. In other words, feature selection is something that needs to be considered because there are a significant number of words that do not contain any information [39]. Naive Bayes (NB) and logistic regression (LR) are two common examples of probabilistic classifiers that can be found among

ML algorithms. The classification of text is the most common application for these algorithms. Also, the NB algorithm makes the fundamental presumption that each feature can be considered independent.

All the aforementioned algorithms can be categorized as nonnetwork-based algorithms. In other words, ML techniques are either based on neural networks or not. Neural networks, also known as NN, have recently been used in a wide variety of applications, including computer systems and artificial intelligence (AI). NN consists of several layers, and these layers can be divided into three types: Input, hidden, and output layers (Figure 14.2). Also, term "Deep Learning" refers to any Neural Networks algorithm with a considerable number of hidden layers [3]. The MLP and the CNN are both examples of multi-layer feedforward networks and nonparametric algorithms that use the NN as part of their ML calculations. Both networks have the ability to recognize images by employing several connected layers. Additional network techniques in the ML are self-organizing map (SOM), RNN, adaptive network-fuzzy interference system (ANFIS), and DBN [3]. As a direct consequence of this, several studies have been conducted in this area.

It is important to note that none of the previously mentioned algorithms is always an ideal solution, and it is highly recommended to select an algorithm based on the dataset type and its features. Besides that, it is possible

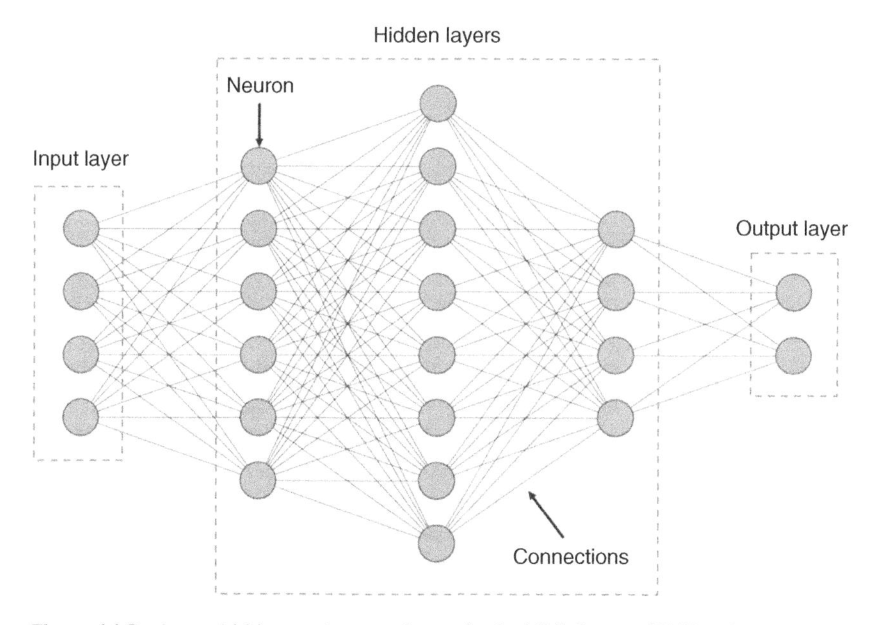

Figure 14.2 Input, hidden, and output layers in the MLP. Source: [3]/Elsevier.

that certain features or attributes of the dataset will have a substantial effect on the prediction. In contrast, a large proportion of a dataset's features might be irrelevant or noisy. Consequently, one should consider the significance of particular features relative to the significance of other features. After that, reliable class correlations are something that needs to be taken into consideration in order to locate the most valuable features of the algorithms.

14.4.3 Application of AI/ML in 3D Printing: A Roadmap from Defect Detection to Optimization Purposes

14.4.3.1 Defect Detection

Although 3D printing is a potential method for manufacturing a variety of 3D structures, various defects may be revealed in the structures produced using this method. In other words, 3D printing is susceptible to having defects and infills due to nonunavoidable factors. Cracks, poor surface quality, and distortion are some of the most common 3D printing defects. Consequently, the product's quality may decline significantly. Not only do defects such as cracks or inconsistency of the extruded filament affect the quality and damage the potential market and customers, but they are also crucial in 3D printing products because defects can cause stress concentration (such as cracks) and affect the mechanical properties. Presumably, these defects weaken the stiffness and yield load, which could lead to unexpected failure and collapse. On the one hand, monitoring and examining the product's quality is likely time-consuming, difficult to handle, and inefficient. In addition, when such monitoring is required, the price of the product increases, and the price increase is not comparable to other techniques on the market. ML can be utilized to address the aforementioned problem.

One of the methods to find defects and stop the printing is real-time defect detection which compares the printed component with the infill pattern owing to the fact that defects are prone to alter the infill pattern [40] (see Figure 14.3). The in situ monitoring system is a robust method in order to find the defect during the process and fix that. By using this system, not only can defects be found, but also it is possible to compensate and fix the defects during the printing process and have a more accurate part. As a result, the time consuming and price does not exceed too much. The two main monitoring systems are machine vision-based and laser scanning-based. In the former monitoring system, after each layer deposition, high-resolution pictures take, and the latter one applies in order to measure the height [40].

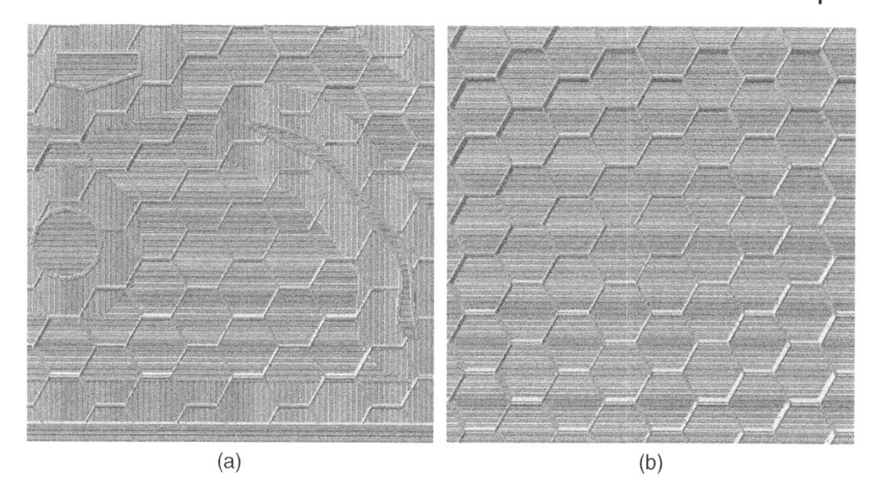

Figure 14.3 Infill patterns (a) with defects (b) without any defects. Source: [40] Khan et al., 2021/with permission from ELSEVIER.

The most important advantage of these monitoring systems is that they can find the defect as well as stop printing and compensate the manufacturing process at the same time. Thus, no additional effort is demanded, and the price does not increase compared to the traditional method, which requires more time to fix that defect and produce waste material. Like two sides of the same coin, this technique has some disadvantages too. As an example, the visual detection method cannot detect internal defects like voids which affect the weight not too much and is hard to detect [41]. Another problem with vision inspection is that different views usually cannot be reachable to analyze the pictures. Accordingly, some of the defects cannot be found.

The main factors associated with the defects include high temperature of the extrusion nozzle, not the appropriate distance between the nozzle tip and printing bed, misalignment of the bed and nozzle, and some miscellaneous factors which can reveal several issues such as bridging and sagging, elephants' foot, overheating and stringing, curling edges, holes, pores, and temperature variation [40]. In order to apply ML for a monitoring system, the training dataset needs to be defined first. Training data can be considered as pictures of models with good (no defect) and bad (having defect(s)) categories. Based on that, the algorithm can train for the main purpose of detecting defects. One of the common approaches for using machine learning in in situ monitoring is using the conventional neural network (CNN) algorithm since it can deal with a plethora of data and images.

14.4.3.2 Processing Parameter Optimization

It is an indisputable fact that additive manufacturing has the potential to produce a wide variety of structures from polymers, metals, and other materials. It is essential to understand the final properties of the structures, which is one of the challenges of this method. Traditionally, these values are determined through experimental and computational optimizations. Although these techniques are useful for processing parameters, they may be time-consuming or even ineffective when faced with a large number of features. Another issue with these methods, particularly FEM, is that the actual properties or behavior of the material may deviate from the predicted properties in light of the fact that some assumptions (such as simplification) have been made, which results in the actual properties. It stands to reason that a more effective strategy should be taken into account.

As an illustration, in the process of laser powder bed fusion, certain physical issues, such as evaporation, melting quickly, recoil, or solidification, can be created and result in porosity or cracks, which makes the component with defects and even inapplicable in some cases. FEM models as well as the high-fidelity models were developed by some researchers in order to get a good grasp of the process and the aforementioned issues [5]. Nevertheless, these models are not applicable in the real world, and their performance may call for a powerful computer.

Among the methods for predicting the behavior of mechanical systems, ML has the advantages of being able to learn from a small number of examples (training set) and carrying out an algorithm that it has learned and there is no need for explicit use [42]. As a result, ML as a data-driven approach can be utilized to accomplish the goals of optimizing parameters and accurately estimating their values.

One of the applications of ML in processing parameter optimization is the laser powder bed fusion in which the remelted depth, based on the features like laser power and scan speed, can be functionalized.

Usually, a probabilistic model is a good choice to predict material behavior as well as mechanical properties [43]. For example, a Gaussian process-based model has been implemented for the prediction of melt pool depth, and scan speed and laser power have been considered the main features of this ML algorithm. Also, MLP, as another ML algorithm, was implemented to predict width, depth, and height in the melt pool [43]. Moreover, MLP is used in order to optimize process parameters, such as layer thickness, build orientation, air gap, and raster angle, and give better viscoelastic properties [43]. Furthermore, SVM was applied in order to study the influence of various factors (features) on the flow mechanism of the cement-based material using 3D printing [43]. One of the applications of unsupervised learning is parameter optimization. As an instance, convolutional autoencoder and clustering have been used for parameter optimization in metal-based AD [44].

Gaussian process as a regression technique attracted more attention for the prediction of the materials. It has been demonstrated that the Gaussian process model has a high degree of precision and can be utilized appropriately in the processing parameter optimization of laser powder bed fusion [8]. Using Gaussian process to construct a regression machine learning algorithm, it is possible to obtain a model that predicts melt pool depth in laser powder bed fusion by using laser power, laser beam size combination, and scan speed as features or inputs [5].

On the microscale, the microstructure of the materials is a crucial factor for the material properties, and as a result, any slight changes affect the properties significantly. Some notable examples of microstructural features include grain size, grain orientation, phase transitions, shape, size, and volume fraction of the particles [43].

14.4.3.3 Geometric Control Using Deep Learning

It is possible to use ML to control the shape of the structures or prototypes that are created through 3D printing. This is made possible by the ability of ML algorithms to detect and correct for deviations and defects. Complex shapes that have a variety of geometries can be very useful in a variety of applications, including industries. In this regard, tolerance and manufacturing of a component with a high standard shape in various industries is one of the challenges that must be overcome. In light of how AM techniques work, factors such as physical connection and chemical reaction (in polymer, for example) can have an impact on the part's final geometry. As a result, the part's final shape and geometry may not be what was expected. On the contrary, it is possible that utilizing some more conventional approaches will not always be practical. For instance, throwing away even a small number of imperfect components is not a good idea as this will result in an increase in the overall cost of production due to the addition of a step to the manufacturing process in order to identify the problematic components. This will also result in the waste of some materials.

For example, laser-based additive manufacturing (LBAM) uses welding to manufacture components with various shapes, some of which are very complex. The main issue of using this technique is related to the product's geometric inaccuracy owing to the fact that the parts experience various temperatures during the manufacturing process which might affects the final shape and even reveals awkward part. In order to rectify this problem, ML was used to adjust the manufacturing path and control the geometry parts.

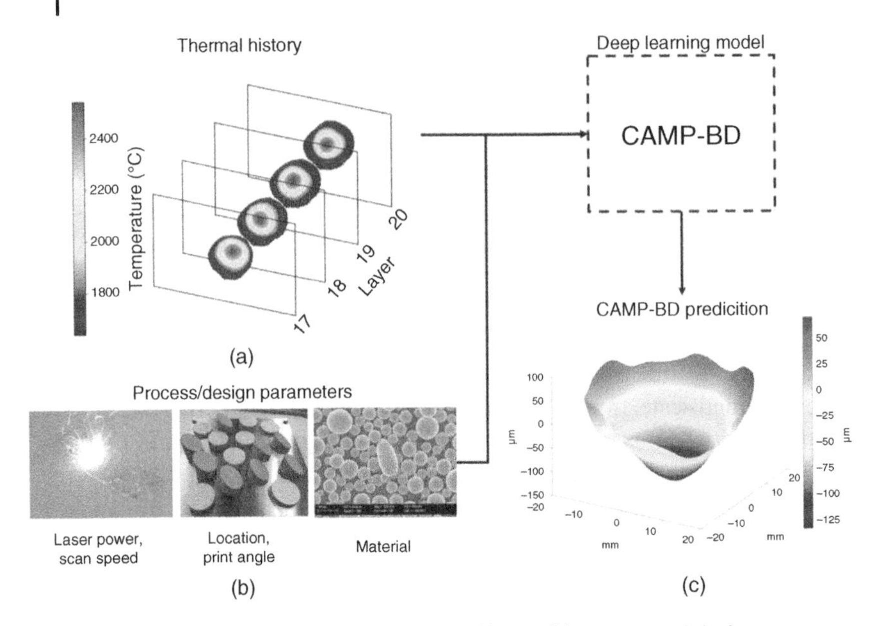

Figure 14.4 CAMP-BD model (a) thermal history, (b) process and design parameters, and (c) the prediction of CAMP-BD. Source: Reproduced with permission from [45]/Elsevier.

In a study by Francis and Bian [45], the authors claimed that deep learning in LBAM is efficient and helpful in increasing the accuracy of the geometry parts, which is very important in Industry 4.0. In order to do that, thermal images from the manufacturing process for the CNN algorithm were used. Also, artificial neural network (ANN) was implemented in order to add process and design parameters. Figure 14.4 shows the fully connected CAMP-BD (convolutional and artificial neural network for additive manufacturing prediction using big data) model. After estimating the distortion, compensation for that distortion can be applied, resulting in a much more accurate geometry (Figure 14.5). Not only can CAMP-BD handle a large dataset, but also this approach is very robust and can be generalized for porosity and residual stress issues [45].

14.4.3.4 Cost Estimation

One more application of machine learning in artificial intelligence is the estimation of the costs of new parts based on previous experience and their similarities to other 3D parts. The traditional method of estimating the cost of 3D printed parts takes into account a number of factors, including the properties of the part, its size, the quality of the machine or printer, and the bidder's level of experience [46]. Using this method does, however, bring

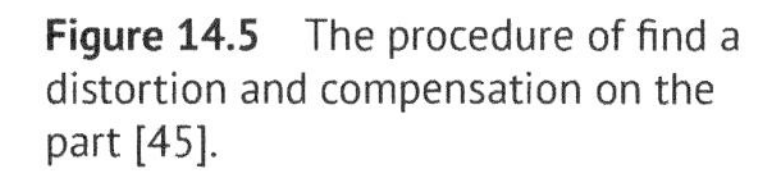

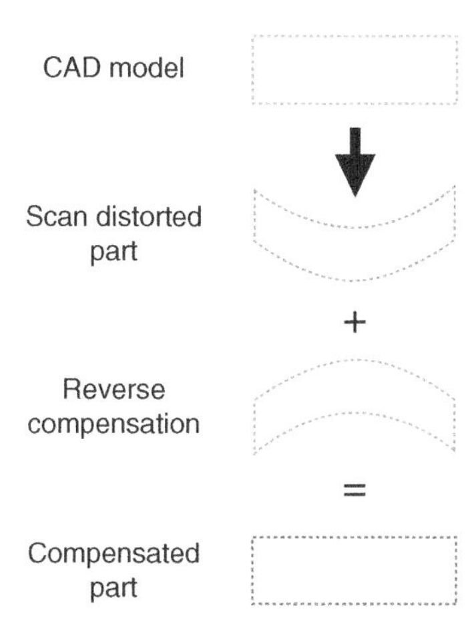

Figure 14.5 The procedure of find a distortion and compensation on the part [45].

up a few issues that need to be addressed. For instance, it does not take into account a great deal of the factors that are associated with the product, and the precision of the price is probably controversial. Thanks to data science and AI, it is possible to use machine learning, which can handle a large dataset or information about the part, and as a result, it can estimate a price in a manner that is significantly more accurate.

As an illustration, the authors in [46] the declare that using a data-driven method can decrease time and the cost estimation process. In order to predict the cost, the statistical ML methods, as well as feature vector construction and dynamic clustering method, were implemented [46]. First, the 3D model with the necessary information, like materials, send to the cost analytic service provider. Second, the features vector defines according to the mentioned information. Then, the training algorithm was implemented on these features (the model has been trained based on similar parts). Additionally, a hierarchical simulation framework can be applied to get a prediction of the manufacturing process. Finally, the cost prediction estimate is based on the prediction of simulation and the cost of past parts.

The integration of machine learning into the realm of 3D printing holds profound significance, ushering in a new era of efficiency, accuracy, and innovation. ML's capacity to analyze vast datasets, learn patterns, and make predictions based on unseen data addresses some of the most pressing challenges within the 3D printing process. By leveraging ML algorithms, the intricate relationship between printing parameters, material properties, and final part quality can be better understood and optimized. The ability to predict and mitigate defects in real time through continuous monitoring enhances the reliability of the printing process, leading to fewer wasted resources and reduced production downtimes. ML's potential in process optimization also extends to tailoring designs for specific printing techniques, minimizing the need for post-processing and thus expediting the overall manufacturing cycle. Moreover, ML-driven generative design opens doors to unparalleled creativity by creating novel geometries that not only push the boundaries of traditional manufacturing but also maximize material utilization and structural integrity. In fields such as healthcare, aerospace, and automotive industries, where precision and innovation are paramount, ML's role in 3D printing transforms how products are conceptualized, designed, and produced. As the synergy between ML and 3D printing advances, it empowers researchers, engineers, and manufacturers to envision new possibilities and overcome historical limitations. The marriage of these two technologies is not merely about optimizing an existing process; it is about catalyzing a paradigm shift in how we imagine, create, and refine objects. To fully harness this potential, continued collaboration between AI specialists and 3D printing experts, along with ongoing research and development, will be essential. In essence, the importance of machine learning in 3D printing is not just in its ability to enhance a technology but to redefine the very essence of what that technology can achieve.

References

1 Goh, G.D., Sing, S.L., and Yeong, W.Y. (2021). A review on machine learning in 3D printing: applications, potential, and challenges. *Artificial Intelligence Review* 54 (1): 63–94.

2 Meng, L., McWilliams, B., Jarosinski, W. et al. (2020). Machine learning in additive manufacturing: a review. *JOM* 72 (6): 2363–2377.

3 Wang, C., Tan, X.P., Tor, S.B., and Lim, C.S. (2020). Machine learning in additive manufacturing: state-of-the-art and perspectives. *Additive Manufacturing* 36: 101538.

4 Zhu, Z., Anwer, N., Huang, Q., and Mathieu, L. (2018). Machine learning in tolerancing for additive manufacturing. *CIRP Annals* 67 (1): 157–160.

5 Tapia, G., Khairallah, S., Matthews, M. et al. (2018). Gaussian process-based surrogate modeling framework for process planning in laser powder-bed fusion additive manufacturing of 316L stainless steel. *The International Journal of Advanced Manufacturing Technology* 94 (9): 3591–3603.

6 Tapia, G., Elwany, A.H., and Sang, H. (2016). Prediction of porosity in metal-based additive manufacturing using spatial Gaussian process models. *Additive Manufacturing* 12: 282–290.

7 Kamath, C. (2016). Data mining and statistical inference in selective laser melting. *The International Journal of Advanced Manufacturing Technology* 86 (5): 1659–1677.
8 Meng, L. and Zhang, J. (2020). Process design of laser powder bed fusion of stainless steel using a Gaussian process-based machine learning model. *JOM* 72 (1): 420–428.
9 Caiazzo, F. and Caggiano, A. (2018). Laser direct metal deposition of 2024 al alloy: trace geometry prediction via machine learning. *Materials* 11 (3): 444.
10 Zhang, J., Wang, P., and Gao, R.X. (2019). Deep learning-based tensile strength prediction in fused deposition modeling. *Computers in Industry* 107: 11–21.
11 Li, Z., Zhang, Z., Shi, J., and Wu, D. (2019). Prediction of surface roughness in extrusion-based additive manufacturing with machine learning. *Robotics and Computer-Integrated Manufacturing* 57: 488–495.
12 Mozaffar, M., Paul, A., Al-Bahrani, R. et al. (2018). Data-driven prediction of the high-dimensional thermal history in directed energy deposition processes via recurrent neural networks. *Manufacturing Letters* 18: 35–39.
13 Song, L., Huang, W., Han, X., and Mazumder, J. (2016). Real-time composition monitoring using support vector regression of laser-induced plasma for laser additive manufacturing. *IEEE Transactions on Industrial Electronics* 64 (1): 633–642.
14 Wang, T., Kwok, T.-H., Zhou, C., and Vader, S. (2018). In-situ droplet inspection and closed-loop control system using machine learning for liquid metal jet printing. *Journal of Manufacturing Systems* 47: 83–92.
15 Yao, X., Moon, S.K., and Bi, G. (2017). A hybrid machine learning approach for additive manufacturing design feature recommendation. *Rapid Prototyping Journal* 23 (6): 983–997. https://doi.org/10.1108/RPJ-03-2016-0041.
16 Zhang, Y., Hong, G.S., Ye, D. et al. (2018). Extraction and evaluation of melt pool, plume and spatter information for powder-bed fusion am process monitoring. *Materials & Design* 156: 458–469.
17 Scime, L. and Beuth, J. (2018). Anomaly detection and classification in a laser powder bed additive manufacturing process using a trained computer vision algorithm. *Additive Manufacturing* 19: 114–126.
18 Khanzadeh, M., Chowdhury, S., Marufuzzaman, M. et al. (2018). Porosity prediction: supervised-learning of thermal history for direct laser deposition. *Journal of Manufacturing Systems* 47: 69–82.
19 Samie Tootooni, M., Dsouza, A., Donovan, R. et al. (2017). Classifying the dimensional variation in additive manufactured parts from

laser-scanned three-dimensional point cloud data using machine learning approaches. *Journal of Manufacturing Science and Engineering* 139 (9): 091005.

20 Aoyagi, K., Wang, H., Sudo, H., and Chiba, A. (2019). Simple method to construct process maps for additive manufacturing using a support vector machine. *Additive Manufacturing* 27: 353–362.

21 Ye, D., Hong, G.S., Zhang, Y. et al. (2018). Defect detection in selective laser melting technology by acoustic signals with deep belief networks. *The International Journal of Advanced Manufacturing Technology* 96 (5): 2791–2801.

22 Shen, Z., Shang, X., Zhao, M. et al. (2019). A learning-based framework for error compensation in 3D printing. *IEEE Transactions on Cybernetics* 49 (11): 4042–4050.

23 Jafari-Marandi, R., Khanzadeh, M., Tian, W. et al. (2019). From in-situ monitoring toward high-throughput process control: cost-driven decision-making framework for laser-based additive manufacturing. *Journal of Manufacturing Systems* 51: 29–41.

24 Gobert, C., Reutzel, E.W., Petrich, J. et al. (2018). Application of supervised machine learning for defect detection during metallic powder bed fusion additive manufacturing using high resolution imaging. *Additive Manufacturing* 21: 517–528.

25 Angelone, R., Caggiano, A., Teti, R. et al. (2020). Bio-intelligent selective laser melting system based on convolutional neural networks for in-process fault identification. *Procedia CIRP* 88: 612–617.

26 Shevchik, S.A., Kenel, C., Leinenbach, C., and Wasmer, K. (2018). Acoustic emission for in situ quality monitoring in additive manufacturing using spectral convolutional neural networks. *Additive Manufacturing* 21: 598–604.

27 Shahrubudin, N., Lee, T.C., and Ramlan, R. (2019). An overview on 3D printing technology: technological, materials, and applications. *Procedia Manufacturing* 35: 1286–1296.

28 Ngo, T.D., Kashani, A., Imbalzano, G. et al. (2018). Additive manufacturing (3D printing): a review of materials, methods, applications and challenges. *Composites Part B: Engineering* 143: 172–196.

29 Talaat, F.M. and Hassan, E. (2021). Artificial intelligence in 3D printing. In: *Enabling Machine Learning Applications in Data Science*, Algorithms for Intelligent Systems (ed. A.E. Hassanien, A. Darwish, S.M. Abd El-Kader, and D.A. Alboaneen), 77–88. Singapore: Springer.

30 Baumers, M. and Ozcan, E. (2016). Scope for machine learning in digital manufacturing. arXiv preprint arXiv:1609.05835.

31 Learned-Miller, E.G. (2014). *Introduction to Supervised Learning. I: Department of Computer Science*, 3. University of Massachusetts.

32 Bousquet, O., von Luxburg, U., and Rätsch, G. (2011). *Advanced Lectures on Machine Learning: ML Summer Schools 2003*, Canberra, Australia, February 2–14, 2003, Tübingen, Germany, August 4–16, 2003, Revised Lectures, vol. 3176. Springer.

33 Jordan, M.I. and Mitchell, T.M. (2015). Machine learning: trends, perspectives, and prospects. *Science* 349 (6245): 255–260.

34 Bonaccorso, G. (2017). *Machine Learning Algorithms*. Packt Publishing Ltd.

35 El Naqa, I. and Murphy, M.J. (2015). What is machine learning? In: *Machine Learning in Radiation Oncology* (ed. I. El Naqa, R. Li, and M. Murphy), 3–11. Cham: Springer.

36 Soofi, A.A. and Awan, A. (2017). Classification techniques in machine learning: applications and issues. *Journal of Basic & Applied Sciences* 13: 459–465.

37 Yildirim, P. (2015). Filter based feature selection methods for prediction of risks in hepatitis disease. *International Journal of Machine Learning and Computing* 5 (4): 258.

38 Mulak, P. and Talhar, N. (2015). Analysis of distance measures using K-nearest neighbor algorithm on KDD dataset. *International Journal of Science and Research* 4 (7): 2319–7064.

39 Feng, G., Guo, J., Jing, B.-Y., and Sun, T. (2015). Feature subset selection using naive Bayes for text classification. *Pattern Recognition Letters* 65: 109–115.

40 Khan, M.F., Alam, A., Siddiqui, M.A. et al. (2021). Real-time defect detection in 3D printing using machine learning. *Materials Today: Proceedings* 42: 521–528.

41 Wu, M., Phoha, V.V., Moon, Y.B., and Belman, A.K. (2016). Detecting malicious defects in 3D printing process using machine learning and image classification. In: *ASME International Mechanical Engineering Congress and Exposition*, vol. 50688, V014T07A004. American Society of Mechanical Engineers.

42 Liu, Y., Zhao, T., Ju, W., and Shi, S. (2017). Materials discovery and design using machine learning. *Journal of Materiomics* 3 (3): 159–177.

43 Nasiri, S. and Khosravani, M.R. (2021). Machine learning in predicting mechanical behavior of additively manufactured parts. *Journal of Materials Research and Technology* 14: 1137–1153.

44 Silbernagel, C., Aremu, A., and Ashcroft, I. (2020). Using machine learning to aid in the parameter optimisation process for metal-based additive manufacturing. *Rapid Prototyping Journal* 26 (4): 625–637.

45 Francis, J. and Bian, L. (2019). Deep learning for distortion prediction in laser-based additive manufacturing using big data. *Manufacturing Letters* 20: 10–14.

46 Chan, S.L., Lu, Y., and Wang, Y. (2018). Data-driven cost estimation for additive manufacturing in cybermanufacturing. *Journal of Manufacturing Systems* 46: 115–126.

Index

Industrial Strategies and Solutions for 3D Printing: Applications and Optimization, First Edition. Edited by Hamid Reza Vanaei, Sofiane Khelladi, and Abbas Tcharkhtchi.

Printed and bound by CPI Group (UK) Ltd, Croydon, CR0 4YY

24/04/2024

14488404-0001